TensorFlow Reinforcement Learning Quick Start Guide

Get up and running with training and deploying intelligent, self-learning agents using Python

Kaushik Balakrishnan

BIRMINGHAM - MUMBAI

TensorFlow Reinforcement Learning Quick Start Guide

Commissioning Editor: Pravin Dandre
Acquisition Editor: Joshua Nadar
Content Development Editor: Roshan Kumar
Technical Editor: Snehal Dalmet
Copy Editor: Safis Editing
Project Coordinator: Namrata Swetta
Proofreader: Safis Editing
Indexer: Pratik Shirodkar
Graphics: Alishon Mendonsa
Production Coordinator: Aparna Bhagat

First published: March 2019

Production reference: 1290319

Published by Packt Publishing Ltd.
Livery Place
35 Livery Street
Birmingham
B3 2PB, UK.

ISBN 978-1-78953-358-3

www.packtpub.com

To Sally, my dearest.

– Kaushik Balakrishnan

`mapt.io`

Mapt is an online digital library that gives you full access to over 5,000 books and videos, as well as industry leading tools to help you plan your personal development and advance your career. For more information, please visit our website.

Why subscribe?

- Spend less time learning and more time coding with practical eBooks and Videos from over 4,000 industry professionals

- Improve your learning with Skill Plans built especially for you

- Get a free eBook or video every month

- Mapt is fully searchable

- Copy and paste, print, and bookmark content

Packt.com

Did you know that Packt offers eBook versions of every book published, with PDF and ePub files available? You can upgrade to the eBook version at `www.packt.com` and as a print book customer, you are entitled to a discount on the eBook copy. Get in touch with us at `customercare@packtpub.com` for more details.

At `www.packt.com`, you can also read a collection of free technical articles, sign up for a range of free newsletters, and receive exclusive discounts and offers on Packt books and eBooks.

Contributors

About the author

Kaushik Balakrishnan works for BMW in Silicon Valley, and applies reinforcement learning, machine learning, and computer vision to solve problems in autonomous driving. Previously, he also worked at Ford Motor Company and NASA Jet Propulsion Laboratory. His primary expertise is in machine learning, computer vision, and high-performance computing, and he has worked on several projects involving both research and industrial applications. He has also worked on numerical simulations of rocket landings on planetary surfaces, and for this he developed several high-fidelity models that run efficiently on supercomputers. He holds a PhD in aerospace engineering from the Georgia Institute of Technology in Atlanta, Georgia.

About the reviewer

Narotam Singh recently took voluntary retirement from his post of meteorologist with the Indian Meteorological Department, Ministry of Earth Sciences, to pursue his dream of learning and helping society. He has been actively involved with various technical programs and the training of GOI officers in the field of IT and communication. He did his masters in the field of electronics, having graduated with a degree in physics. He also holds a diploma and a postgraduate diploma in the field of computer engineering. Presently, he works as a freelancer. He has many research publications to his name and has also served as a technical reviewer for numerous books. His present research interests involve AI, ML, DL, robotics, and spirituality.

Packt is searching for authors like you

If you're interested in becoming an author for Packt, please visit `authors.packtpub.com` and apply today. We have worked with thousands of developers and tech professionals, just like you, to help them share their insight with the global tech community. You can make a general application, apply for a specific hot topic that we are recruiting an author for, or submit your own idea.

Table of Contents

Preface

This book provides a summary of several different **reinforcement learning** (**RL**) algorithms, including the theory involved in the algorithms as well as coding them using Python and TensorFlow. Specifically, the algorithms covered in this book are Q-learning, SARSA, DQN, DDPG, A3C, TRPO, and PPO. The applications of these RL algorithms include computer games from OpenAI Gym and autonomous driving using the TORCS racing car simulator.

Who this book is for

This book is designed for **machine learning** (**ML**) practitioners interested in learning RL. It will help ML engineers, data scientists, and graduate students. A basic knowledge of ML, and experience of coding in Python and TensorFlow, is expected of the reader in order to be able to complete this book successfully.

What this book covers

Chapter 1, *Up and Running with Reinforcement Learning*, provides an overview of the basic concepts of RL, such as an agent, an environment, and the relationship between them. It also covers topics such as reward functions, discounted rewards, and value and advantage functions. The reader will also get familiar with the Bellman equation, on-policy and off-policy algorithms, as well as model-free and model-based RL algorithms.

Chapter 2, *Temporal Difference, SARSA, and Q-learning*, introduces the reader to temporal difference learning, SARSA, and Q-learning. It also summarizes how to code these algorithms in Python, and to train and test them on two classical RL problems – GridWorld and Cliff Walking.

Chapter 3, *Deep Q-Network*, introduces the reader to the first deep RL algorithm of the book, DQN. It will also discuss how to code this in Python and TensorFlow. The code will then be used to train an RL agent to play *Atari Breakout*.

Chapter 4, *Double DQN, Dueling Architectures, and Rainbow*, builds on the previous chapter and extends it to double DQN. It also discusses dueling network architectures that involve value and advantage streams. These extensions will be coded in Python and TensorFlow, and will be used to train RL agents to play *Atari Breakout*. Finally, Google's dopamine code will be introduced and will be used to train a Rainbow DQN agent.

Chapter 5, *Deep Deterministic Policy Gradient*, is the first actor-critic algorithm of the book as well as the first RL algorithm for continuous control. It introduces policy gradients to the reader and discusses how to use it to train the policy for the actor. The chapter will code this algorithm using Python and TensorFlow and use it to train an agent to play the inverted pendulum problem.

Chapter 6, *Asynchronous Methods – A3C and A2C*, introduces the reader to the A3C algorithm, which is an asynchronous RL algorithm where one master processor will update the policy network, and multiple worker processors will use it to collect experience samples, which will be used to compute the policy gradients, and then passed on to the master processor. Also in this chapter, A3C will be used to train RL agents to play OpenAI Gym's *CartPole* and *LunarLander*. Finally, A2C is also briefly introduced.

Chapter 7, *Trust Region Policy Optimization and Proximal Policy Optimization*, discusses two RL algorithms based on the policy distribution ratio—TRPO and PPO. This chapter also discusses how to code PPO using Python and TensorFlow, and will use it to train an RL agent to solve the MountainCar problem in OpenAI Gym.

Chapter 8, *Deep RL Applied to Autonomous Driving*, introduces the reader to the TORCS racing car simulator, coding the DDPG algorithm for training an agent to drive a car autonomously. The code files for this chapter also include the PPO algorithm for the same TORCS problem, and is provided as an exercise for the reader.

To get the most out of this book

The reader is expected to have a good knowledge of ML algorithms, such as deep neural networks, convolutional neural networks, stochastic gradient descent, and Adam optimization. The reader is also expected to have hands-on coding experience in Python and TensorFlow.

Download the example code files

You can download the example code files for this book from your account at www.packt.com. If you purchased this book elsewhere, you can visit www.packt.com/support and register to have the files emailed directly to you.

You can download the code files by following these steps:

1. Log in or register at www.packt.com.
2. Select the **SUPPORT** tab.
3. Click on **Code Downloads & Errata**.
4. Enter the name of the book in the **Search** box and follow the onscreen instructions.

Once the file is downloaded, please make sure that you unzip or extract the folder using the latest version of:

- WinRAR/7-Zip for Windows
- Zipeg/iZip/UnRarX for Mac
- 7-Zip/PeaZip for Linux

The code bundle for the book is also hosted on GitHub at https://github.com/PacktPublishing/TensorFlow-Reinforcement-Learning-Quick-Start-Guide. In case there's an update to the code, it will be updated on the existing GitHub repository.

We also have other code bundles from our rich catalog of books and videos available at https://github.com/PacktPublishing/. Check them out!

Download the color images

We also provide a PDF file that has color images of the screenshots/diagrams used in this book. You can download it here: http://www.packtpub.com/sites/default/files/downloads/9781789533583_ColorImages.pdf.

Conventions used

There are a number of text conventions used throughout this book.

CodeInText: Indicates code words in text, database table names, folder names, filenames, file extensions, pathnames, dummy URLs, user input, and Twitter handles. Here is an example: "Mount the downloaded WebStorm-10*.dmg disk image file as another disk in your system."

A block of code is set as follows:

```
import numpy as np
import sys
import matplotlib.pyplot as plt
```

When we wish to draw your attention to a particular part of a code block, the relevant lines or items are set in bold:

```
def random_action():
    # a = 0 : top/north
    # a = 1 : right/east
    # a = 2 : bottom/south
    # a = 3 : left/west
    a = np.random.randint(nact)
    return a
```

Any command-line input or output is written as follows:

```
sudo apt-get install python-numpy python-scipy python-matplotlib
```

Bold: Indicates a new term, an important word, or words that you see on screen. For example, words in menus or dialog boxes appear in the text like this. Here is an example: "Select **System info** from the **Administration** panel."

Warnings or important notes appear like this.

Tips and tricks appear like this.

Get in touch

Feedback from our readers is always welcome.

General feedback: If you have questions about any aspect of this book, mention the book title in the subject of your message and email us at customercare@packtpub.com.

Errata: Although we have taken every care to ensure the accuracy of our content, mistakes do happen. If you have found a mistake in this book, we would be grateful if you would report this to us. Please visit www.packt.com/submit-errata, selecting your book, clicking on the Errata Submission Form link, and entering the details.

Piracy: If you come across any illegal copies of our works in any form on the internet, we would be grateful if you would provide us with the location address or website name. Please contact us at copyright@packt.com with a link to the material.

If you are interested in becoming an author: If there is a topic that you have expertise in, and you are interested in either writing or contributing to a book, please visit authors.packtpub.com.

Reviews

Please leave a review. Once you have read and used this book, why not leave a review on the site that you purchased it from? Potential readers can then see and use your unbiased opinion to make purchase decisions, we at Packt can understand what you think about our products, and our authors can see your feedback on their book. Thank you!

For more information about Packt, please visit packt.com.

1
Up and Running with Reinforcement Learning

This book will cover interesting topics in deep **Reinforcement Learning** (**RL**), including the more widely used algorithms, and will also provide TensorFlow code to solve many challenging problems using deep RL algorithms. Some basic knowledge of RL will help you pick up the advanced topics covered in this book, but the topics will be explained in a simple language that machine learning practitioners can grasp. The language of choice for this book is Python, and the deep learning framework used is TensorFlow, and we expect you to have a reasonable understanding of the two. If not, there are several Packt books that cover these topics. We will cover several different RL algorithms, such as **Deep Q-Network** (**DQN**), **Deep Deterministic Policy Gradient** (**DDPG**), **Trust Region Policy Optimization** (**TRPO**), and **Proximal Policy Optimization** (**PPO**), to name a few. Let's dive right into deep RL.

In this chapter, we will delve deep into the basic concepts of RL. We will learn the meaning of the RL jargon, the mathematical relationships between them, and also how to use them in an RL setting to train an agent. These concepts will lay the foundations for us to learn RL algorithms in later chapters, along with how to apply them to train agents. Happy learning!

Some of the main topics that will be covered in this chapter are as follows:

- Formulating the RL problem
- Understanding what an agent and an environment are
- Defining the Bellman equation
- On-policy versus off-policy learning
- Model-free versus model-based training

Why RL?

RL is a sub-field of machine learning where the learning is carried out by a trial-and-error approach. This differs from other machine learning strategies, such as the following:

- **Supervised learning**: Where the goal is to learn to fit a model distribution that captures a given labeled data distribution
- **Unsupervised learning**: Where the goal is to find inherent patterns in a given dataset, such as clustering

RL is a powerful learning approach, since you do not require labeled data, provided, of course, that you can master the learning-by-exploration approach used in RL.

While RL has been around for over three decades, the field has gained a new resurgence in recent years with the successful demonstration of the use of deep learning in RL to solve real-world tasks, wherein deep neural networks are used to make decisions. The coupling of RL with deep learning is typically referred to as deep RL, and is the main topic of discussion of this book.

Deep RL has been successfully applied by researchers to play video games, to drive cars autonomously, for industrial robots to pick up objects, for traders to make portfolio bets, by healthcare practitioners, and copious other examples. Recently, Google DeepMind built AlphaGo, a RL-based system that was able to play the game Go, and beat the champions of the game easily. OpenAI built another system to beat humans in the Dota video game. These examples demonstrate the real-world applications of RL. It is widely believed that this field has a very promising future, since you can train neural networks to make predictions without providing labeled data.

Now, let's delve into the formulation of the RL problem. We will compare how RL is similar in spirit to a child learning to walk.

Formulating the RL problem

The basic problem that is solved is training a model to make predictions of some pre-defined task without any labeled data. This is accomplished by a trial-and-error approach, akin to a baby learning to walk for the first time. A baby, curious to explore the world around them, first crawls out of their crib not knowing where to go nor what to do. Initially, they take small steps, make mistakes, keep falling on the floor, and cry. But, after many such episodes, they start to stand on their feet on their own, much to the delight of their parents. Then, with a giant leap of faith, they start to take slightly longer steps, slowly and cautiously. They still make mistakes, albeit fewer than before.

After many more such tries—and failures—they gain more confidence that enables them to take even longer steps. With time, these steps get much longer and faster, until eventually, they start to run. And that's how they grow up into a child. Was any labeled data provided to them that they used to learn to walk? No. they learned by trial and error, making mistakes along the way, learning from them, and getting better at it with infinitesimal gains made for every attempt. This is how RL works, learning by trial and error.

Building on the preceding example, here is another situation. Suppose you need to train a robot using trial and error, this is how to do it. Let the robot wander randomly in the environment initially. The good and bad actions are collected and a reward function is used to quantify them, thus, a good action performed in a state will have high rewards; on the other hand, bad actions will be penalized. This can be used as a learning signal for the robot to improve itself. After many such episodes of trial and error, the robot would have learned the best action to perform in a given state, based on the reward. This is how learning in RL works. But we will not talk about human characters for the rest of the book. The child described previously is the *agent,* and their surroundings are the *environment* in RL parlance. The agent interacts with the environment and, in the process, learns to undertake a task, for which the environment will provide a reward.

The relationship between an agent and its environment

At a very basic level, RL involves an agent and an environment. An agent is an artificial intelligence entity that has certain goals, must remain vigilant about things that can come in the way of these goals, and must, at the same time, pursue the things that help in the attaining of these goals. An environment is everything that the agent can interact with. Let me explain further with an example that involves an industrial mobile robot.

For example, in a setting involving an industrial mobile robot navigating inside a factory, the robot is the agent, and the factory is the environment.

The robot has certain pre-defined goals, for example, to move goods from one side of the factory to the other without colliding with obstacles such as walls and/or other robots. The environment is the region available for the robot to navigate and includes all the places the robot can go to, including the obstacles that the robot could crash in to. So the primary task of the robot, or more precisely, the agent, is to explore the environment, understand how the actions it takes affects its rewards, be cognizant of the obstacles that can cause catastrophic crashes or failures, and then master the art of maximizing the goals and improving its performance over time.

In this process, the agent inevitably interacts with the environment, which can be good for the agent regarding certain tasks, but could be bad for the agent regarding other tasks. So, the agent must learn how the environment will respond to the actions that are taken. This is a trial-and-error learning approach, and only after numerous such trials can the agent learn how the environment will respond to its decisions.

Let's now come to understand what the state space of an agent is, and the actions that the agent performs to explore the environment.

Defining the states of the agent

In RL parlance, *states* represent the current situation of the agent. For example, in the previous industrial mobile robot agent case, the state at a given time instant is the location of the robot inside the factory – that is, where it is located, its orientation, or more precisely, the pose of the robot. For a robot that has joints and effectors, the state can also include the precise location of the joints and effectors in a three-dimensional space. For an autonomous car, its state can represent its speed, location on a map, distance to other obstacles, torques on its wheels, the rpm of the engine, and so on.

States are usually deduced from sensors in the real world; for instance, the measurement from odometers, LIDARs, radars, and cameras. States can be a one-dimensional vector of real numbers or integers, or two-dimensional camera images, or even higher-dimensional, for instance, three-dimensional voxels. There are really no precise limitations on states, and the state just represents the current situation of the agent.

In RL literature, states are typically represented as s_t, where the subscript t is used to denote the time instant corresponding to the state.

Defining the actions of the agent

The agent performs actions to explore the environment. Obtaining this action vector is the primary goal in RL. Ideally, you need to strive to obtain optimal actions.

An action is the decision an agent takes in a certain state, s_t. Typically, it is represented as a_t, where, as before, the subscript t denotes the time instant. The actions that are available to an agent depends on the problem. For instance, an agent in a maze can decide to take a step north, or south, or east, or west. These are called **discrete actions**, as there are a fixed number of possibilities. On the other hand, for an autonomous car, actions can be the steering angle, throttle value, brake value, and so on, which are called **continuous actions** as they can take real number values in a bounded range. For example, the steering angle can be 40 degrees from the north-south line, and the throttle can be 60% down, and so on.

Thus, actions a_t can be either discrete or continuous, depending on the problem at hand. Some RL approaches handle discrete actions, while others are suited for continuous actions.

A schematic of the **agent** and its interaction with the **environment** is shown in the following diagram:

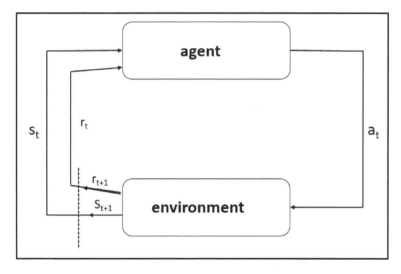

Figure 1: Schematic showing the agent and its interaction with the environment

Now that we know what an **agent** is, we will look at the policies that the agent learns, what value and advantage functions are, and how these quantities are used in RL.

Understanding policy, value, and advantage functions

A **policy** defines the guidelines for an agent's behavior at a given state. In mathematical terms, a policy is a mapping from a state of the agent to the action to be taken at that state. It is like a stimulus-response rule that the agent follows as it learns to explore the environment. In RL literature, it is usually denoted as $\pi(a_t|s_t)$ – that is, it is a conditional probability distribution of taking an action a_t in a given state s_t. Policies can be deterministic, wherein the exact value of a_t is known at s_t, or can be stochastic where a_t is sampled from a distribution – typically this is a Gaussian distribution, but it can also be any other probability distribution.

In RL, **value functions** are used to define how good a state of an agent is. They are typically denoted by $V(s)$ at state s and represent the expected long-term average rewards for being in that state. $V(s)$ is given by the following expression where $E[.]$ is an expectation over samples:

$$V(s) = E[R_t|s_t = s] = E[\sum_{k=0}^{T} \gamma^k r_{t+k+1}|s_t = s]$$

Note that $V(s)$ does not care about the optimum actions that an agent needs to take at the state s. Instead, it is a measure of how good a state is. So, how can an agent figure out the most optimum action a_t to take in a given state s_t at time instant t? For this, you can also define an action-value function given by the following expression:

$$Q(s,a) = E[R_t|s_t = s, a_t = a] = E[\sum_{k=0}^{T} \gamma^k r_{t+k+1}|s_t = s, a_t = a]$$

Note that $Q(s,a)$ is a measure of how good is it to take action a in state s and follow the same policy thereafter. So, t is different from $V(s)$, which is a measure of how good a given state is. We will see in the following chapters how the value function is used to train the agent under the RL setting.

The **advantage function** is defined as the following:

$$A(s,a) = Q(s,a) - V(s)$$

This advantage function is known to reduce the variance of policy gradients, a topic that will be discussed in depth in a later chapter.

The classic RL textbook is *Reinforcement Learning: An Introduction* by *Richard S Sutton* and *Andrew G Barto, The MIT Press*, 1998.

We will now define what an episode is and its significance in an RL context.

Identifying episodes

We mentioned earlier that the agent explores the environment in numerous trials-and-errors before it can learn to maximize its goals. Each such trial from start to finish is called an **episode**. The start location may or may not always be from the same location. Likewise, the finish or end of the episode can be a happy or sad ending.

A happy, or good, ending can be when the agent accomplishes its pre-defined goal, which could be successfully navigating to a final destination for a mobile robot, or successfully picking up a peg and placing it in a hole for an industrial robot arm, and so on. Episodes can also have a sad ending, where the agent crashes into obstacles or gets trapped in a maze, unable to get out of it, and so on.

In many RL problems, an upper bound in the form of a fixed number of time steps is generally specified for terminating an episode, although in others, no such bound exists and the episode can last for a very long time, ending with the accomplishment of a goal or by crashing into obstacles or falling off a cliff, or something similar. The Voyager spacecraft was launched by NASA in 1977, and has traveled outside our solar system – this is an example of a system with an infinite time episode.

We will next find out what a reward function is and why we need to discount future rewards. This reward function is the key, as it is the signal for the agent to learn.

Identifying reward functions and the concept of discounted rewards

Rewards in RL are no different from real world rewards – we all receive good rewards for doing well, and bad rewards (aka penalties) for inferior performance. Reward functions are provided by the environment to guide an agent to learn as it explores the environment. Specifically, it is a measure of how well the agent is performing.

The reward function defines what the good and bad things are that can happen to the agent. For instance, a mobile robot that reaches its goal is rewarded, but is penalized for crashing into obstacles. Likewise, an industrial robot arm is rewarded for putting a peg into a hole, but is penalized for being in undesired poses that can be catastrophic by causing ruptures or crashes. Reward functions are the signal to the agent regarding what is optimum and what isn't. The agent's long-term goal is to maximize rewards and minimize penalties.

Rewards

In RL literature, rewards at a time instant t are typically denoted as R_t. Thus, the total rewards earned in an episode is given by $R = r1 + r2 + ... + r_t$, where t is the length of the episode (which can be finite or infinite).

The concept of discounting is used in RL, where a parameter called the discount factor is used, typically represented by γ and $0 \leq \gamma \leq 1$ and a power of it multiplies r_t. $\gamma = 0$, making the agent myopic, and aiming only for the immediate rewards. $\gamma = 1$ makes the agent far-sighted to the point that it procrastinates the accomplishment of the final goal. Thus, a value of γ in the 0-1 range (0 and 1 exclusive) is used to ensure that the agent is neither too myopic nor too far-sighted. γ ensures that the agent prioritizes its actions to maximize the total discounted rewards, R_t, from time instant t, which is given by the following:

$$R_t = \sum_{k=t}^{T} \gamma^{(k-t)} r_k(s_k, a_k)$$

Since $0 \leq \gamma \leq 1$, the rewards into the distant future are valued much less than the rewards that the agent can earn in the immediate future. This helps the agent to not waste time and to prioritize its actions. In practice, $\gamma = 0.9\text{-}0.99$ is typically used in most RL problems.

Learning the Markov decision process

The Markov property is widely used in RL, and it states that the environment's response at time $t+1$ depends only on the state and action at time t. In other words, the immediate future only depends on the present and not on the past. This is a useful property that simplifies the math considerably, and is widely used in many fields such as RL and robotics.

Consider a system that transitions from state s_0 to s_1 by taking an action a_0 and receiving a reward r_1, and thereafter from s_1 to s_2 taking action a_1, and so on until time t. If the probability of being in a state s' at time $t+1$ can be represented mathematically as in the following function, then the system is said to follow the Markov property:

$$Pr(s_{t+1} = s', r_{t+1} = r | s_t, a_t, r_t, s_{t-1}, a_{t-1}, \ldots, r_1, s_0, a_0) = Pr(s_{t+1} = s', r_{t+1} = r | s_t, a_t)$$

Note that the probability of being in state s_{t+1} depends only on s_t and a_t and not on the past. An environment that satisfies the following state transition property and reward function as follows is said to be a **Markov Decision Process (MDP)**:

$$P_{ss'} = Pr[s_{t+1} = s' | s_t = s, a_t = a]$$

$$R_{ss'} = R[r_{t+1} | s_t = s, a_t = a, s_{t+1} = s']$$

Let's now define the very foundation of RL: the Bellman equation. This equation will help in providing an iterative solution to obtaining value functions.

Defining the Bellman equation

The Bellman equation, named after the great computer scientist and applied mathematician Richard E. Bellman, is an optimality condition associated with dynamic programming. It is widely used in RL to update the policy of an agent.

Let's define the following two quantities:

$$P_{s,s'} = Pr(s_{t+1} = s' | s_t = s, a_t = a)$$

$$R_{s,s'} = E[r_{t+1} | s_t = s, s_{t+1} = s', a_t = a]$$

The first quantity, $P_{s,s'}$, is the transition probability from state s to the new state s'. The second quantity, $R_{s,s'}$, is the expected reward the agent receives from state s, taking action a, and moving to the new state s'. Note that we have assumed the MDP property, that is, the transition to state at time *t+1* only depends on the state and action at time *t*. Stated in these terms, the Bellman equation is a recursive relationship, and is given by the following equations respectively for the value function and action-value function:

$$V(s) = \sum_{a} \pi(s, a) \sum_{s'} P_{ss'} [R_{ss'} + \gamma V(s')]$$

$$Q(s, a) = \sum_{s'} P_{ss'} [R_{ss'} + \gamma \sum_{a'} \pi(s', a') Q(s', a')]$$

Note that the Bellman equations represent the value function V at a state, and as functions of the value function at other states; similarly for the action-value function Q.

On-policy versus off-policy learning

RL algorithms can be classified as on-policy or off-policy. We will now learn about both of these classes and how to distinguish a given RL algorithm into one or the other.

On-policy method

On-policy methods use the same policy to evaluate as was used to make the decisions on actions. On-policy algorithms generally do not have a replay buffer; the experience encountered is used to train the model in situ. The same policy that was used to move the agent from state at time *t* to state at time *t+1*, is used to evaluate if the performance was good or bad. For example, if a robot is exploring the world at a given state, if it uses its current policy to ascertain whether the actions it took in the current state were good or bad, then it is an on-policy algorithm, as the current policy is also used to evaluate its actions. SARSA, A3C, TRPO, and PPO are on-policy algorithms that we will be covering in this book.

Off-policy method

Off-policy methods, on the other hand, use different policies to make action decisions and to evaluate the performance. For instance, many off-policy algorithms use a replay buffer to store the experiences, and sample data from this buffer to train the model. During the training step, a mini-batch of experience data is randomly sampled and used to train the policy and value functions. Coming back to the previous robot example, in an off-policy setting, the robot will not use the current policy to evaluate its performance, but rather use a different policy for exploring and for evaluation. If a replay buffer is used to sample a mini-batch of experience data and then train the agent, then it is off-policy learning, as the current policy of the robot (which was used to obtain the immediate actions) is different from the policy that was used to obtain the samples in the mini-batch of experience used to train the agent (as the policy has changed from an earlier time instant when the data was collected, to the current time instant). DQN, DDQN, and DDPG are off-policy algorithms that we'll look at in later chapters of this book.

Model-free and model-based training

RL algorithms that do not learn a model of how the environment works are called model-free algorithms. By contrast, if a model of the environment is constructed, then the algorithm is called model-based. In general, if value (V) or action-value (Q) functions are used to evaluate the performance, they are called **model-free** algorithms as no specific model of the environment is used. On the other hand, if you build a model of how the environment transitions from one state to another or determines how many rewards the agent will receive from the environment via a model, then they are called **model-based** algorithms.

In model-free algorithms, as aforementioned, we do not construct a model of the environment. Thus, the agent has to take an action at a state to figure out if it is a good or a bad choice. In model-based RL, an approximate model of the environment is learned; either jointly learned along with the policy, or learned a priori. This model of the environment is used to make decisions, as well as to train the policy. We will learn more about both classes of RL algorithms in later chapters.

Algorithms covered in this book

In Chapter 2, *Temporal Difference, SARSA, and Q-Learning*, we will look into our first two RL algorithms: Q-learning and SARSA. Both of these algorithms are tabular-based and do not require the use of neural networks. Thus, we will code them in Python and NumPy. In Chapter 3, *Deep Q-Network*, we will cover DQN and use TensorFlow to code the agent for the rest of the book. We will then train it to play Atari Breakout. In Chapter 4, *Double DQN, Dueling Architectures, and Rainbow*, we will cover double DQN, dueling network architectures, and rainbow DQN. In Chapter 5, *Deep Deterministic Policy Gradient*, we will look at our first Actor-Critic RL algorithm called DDPG, learn about policy gradients, and apply them to a continuous action problem. In Chapter 6, *Asynchronous Methods – A3C and A2C*, we will investigate A3C, which is another RL algorithm that uses a master and several worker processes. In Chapter 7, *Trust Region Policy Optimization and Proximal Policy Optimization*, we will investigate two more RL algorithms: TRPO and PPO. Finally, we will apply DDPG and PPO to train an agent to drive a car autonomously in Chapter 8, *Deep RL Applied to Autonomous Driving*. From Chapter 3, *Deep Q-Network*, to Chapter 8, *Deep RL Applied to Autonomous Driving*, we'll use TensorFlow agents. Have fun learning RL.

Summary

In this chapter, we were introduced to the basic concepts of RL. We understood the relationship between an agent and its environment, and also learned about the MDP setting. We learned the concept of reward functions and the use of discounted rewards, as well as the idea of value and advantage functions. In addition, we saw the Bellman equation and how it is used in RL. We also learned the difference between an on-policy and an off-policy RL algorithm. Furthermore, we examined the distinction between model-free and model-based RL algorithms. All of this lays the groundwork for us to delve deeper into RL algorithms and how we can use them to train agents for a given task.

In the next chapter, we will investigate our first two RL algorithms: Q-learning and SARSA. Note that in Chapter 2, *Temporal Difference, SARSA, and Q-Learning*, we will be using Python-based agents as they are tabular-learning. But from Chapter 3, *Deep Q-Network*, onward, we will be using TensorFlow to code deep RL agents, as we will require neural networks.

Questions

1. Is a replay buffer required for on-policy or off-policy RL algorithms?
2. Why do we discount rewards?
3. What will happen if the discount factor is $\gamma > 1$?
4. Will a model-based RL agent always perform better than a model-free RL agent, since we have a model of the environment states?
5. What is the difference between RL and deep RL?

Further reading

- *Reinforcement Learning: An Introduction, Richard S. Sutton* and *Andrew G. Barto, The MIT Press*, 1998
- *Deep Reinforcement Learning Hands-On, Maxim Lapan, Packt Publishing,* 2018: https://www.packtpub.com/big-data-and-business-intelligence/deep-reinforcement-learning-hands

Temporal Difference, SARSA, and Q-Learning

2

In the previous chapter, we looked at the basics of RL. In this chapter, we will cover **temporal difference** (**TD**) learning, SARSA, and Q-learning, which were very widely used algorithms in RL before deep RL became more common. Understanding these older-generation algorithms is essential if you want to master the field, and will also lay the foundation for delving into deep RL. We will therefore spend this chapter looking at examples using these older generation algorithms. In addition, we will also code some of these algorithms using Python. We will not be using TensorFlow for this chapter, as the problems do not involve any deep neural networks under study. However, this chapter will lay the groundwork for more advanced topics that we will cover in the subsequent chapters, and will also be our first coding experience of an RL algorithm. Specifically, this chapter will be our first deep dive into a standard RL algorithm, and how you can use it to train RL agents for a specific task. It will also be our first hands-on effort at RL, including both theory and practice.

Some of the topics that will be covered in this chapter are as follows:

- Understanding TD learning
- Learning SARSA
- Understanding Q-learning
- Cliff walking with SARSA and Q-learning
- Grid world with SARSA

Technical requirements

Knowledge of the following will help you to better understand the concepts presented in this chapter:

- Python (version 2 or 3)
- NumPy
- TensorFlow (version 1.4 or higher)

Understanding TD learning

We will first learn about TD learning. This is a very fundamental concept in RL. In TD learning, the learning of the agent is attained by experience. Several trial episodes are undertaken of the environment, and the rewards accrued are used to update the value functions. Specifically, the agent will keep an update of the state-action value functions as it experiences new states/actions. The Bellman equation is used to update this state-action value function, and the goal is to minimize the TD error. This essentially means the agent is reducing its uncertainty of which action is the optimal action in a given state; it gains confidence on the optimal action in a given state by lowering the TD error.

Relation between the value functions and state

The value function is an agent's estimate of how good a given state is. For instance, if a robot is near the edge of a cliff and may fall, that state is bad and must have a low value. On the other hand, if the robot/agent is near its final goal, that state is a good state to be in, as the rewards they will soon receive are high, and so that state will have a higher value.

The value function, V, is updated after reaching a s_t state and receiving a r_t reward from the environment. The simplest TD learning algorithm is called *TD(0)* and performs an update using the following equation where α is the learning rate and $0 \leq \alpha \leq 1$:

$$V(s_t) := V(s_t) + \alpha[r_{t+1} + \gamma V(s_{t+1}) - V(s_t)]$$

Note that in some reference papers or books, the preceding formula will have r_t instead of r_{t+1}. This is just a difference in convention and is not an error; r_{t+1} here denotes the reward received from s_t state and transitioning to s_{t+1}.

There is also another TD learning variant called *TD(λ)* that used eligibility traces *e(s)*, which are a record of visiting a state. More formally, we perform a *TD(λ)* update as follows:

$$V(s_t) := V(s_t) + \alpha[r_{t+1} + \gamma V(s_{t+1}) - V(s_t)]e(s_t)$$

The eligibility traces are given by the following equation:

$$e(s_t) = \begin{cases} \gamma\lambda e(s_{t-1}) & \text{if } s \neq s_t \\ \gamma\lambda e(s_{t-1}) + 1 & \text{if } s = s_t \end{cases}$$

Here, *e(s)* = 0 at *t* = 0. For each step the agent takes, the eligibility trace decreases by $\gamma\lambda$ for all states, and is incremented by *1* for the state visited in the current time step. Here, $0 \leq \lambda \leq 1$, and it is a parameter that decides how much of the credit from a reward is to be assigned to distant states. Next, we will look at the theory behind our next two RL algorithms, SARSA and Q-learning, both of which are quite popular in the RL community.

Understanding SARSA and Q-Learning

In this section, we will learn about SARSA and Q-Learning and how can they are coded with Python. Before we go further, let's find out what SARSA and Q-Learning are. SARSA is an algorithm that uses the state-action Q values to update. These concepts are derived from the computer science field of dynamic programming, while Q-learning is an off-policy algorithm that was first proposed by Christopher Watkins in 1989, and is a widely used RL algorithm.

Learning SARSA

SARSA is another on-policy algorithm that was very popular, particularly in the 1990s. It is an extension of TD-learning, which we saw previously, and is an on-policy algorithm. SARSA keeps an update of the state-action value function, and as new experiences are encountered, this state-action value function is updated using the Bellman equation of dynamic programming. We extend the preceding TD algorithm to state-action value function, $Q(s_t, a_t)$, and this approach is called SARSA:

$$Q(s_t, a_t) := Q(s_t, a_t) + \alpha[r_{t+1} + \gamma Q(s_{t+1}, a_{t+1}) - Q(s_t, a_t)]$$

Here, from a given state s_t, we take action a_t, receive a reward r_{t+1}, transition to a new state s_{t+1}, and thereafter take an action a_{t+1} that then continues on and on. This quintuple (s_t, a_t, r_{t+1}, s_{t+1}, a_{t+1}) gives the algorithm the name SARSA. SARSA is an on-policy algorithm, as the same policy is updated as was used to estimate Q. For exploration, you can use, say, ε-greedy.

Understanding Q-learning

Q-learning is an off-policy algorithm that was first proposed by Christopher Watkins in 1989, and is a widely used RL algorithm. Q-learning, such as SARSA, keeps an update of the state-action value function for each state-action pair, and recursively updates it using the Bellman equation of dynamic programming as new experiences are collected. Note that it is an off-policy algorithm as it uses the state-action value function evaluated at the action, which will maximize the value. Q-learning is used for problems where the actions are discrete – for example, if we have the actions move north, move south, move east, move west, and we are to decide the optimum action in a given state, then Q-learning is applicable in such settings.

In the classical Q-learning approach, the update is given as follows, where the max is performed over actions, that is, we choose the action a corresponding to the maximum value of Q at state s_{t+1}:

$$Q(s_t, a_t) := Q(s_t, a_t) + \alpha[r_{t+1} + \gamma \ \max_a \ Q(s_{t+1}, a) - Q(s_t, a_t)]$$

The α is the learning rate, which is a hyper-parameter that the user can specify.

Before we code the algorithms in Python, let's find out what kind of problems will be considered.

Cliff walking and grid world problems

Let's consider cliff walking and grid world problems. First, we will introduce these problems to you, then we will proceed on to the coding part. For both problems, we consider a rectangular grid with `nrows` (number of rows) and `ncols` (number of columns). We start from one cell to the south of the bottom left cell, and the goal is to reach the destination, which is one cell to the south of the bottom right cell.

Note that the start and destination cells are not part of the `nrows` x `ncols` grid of cells. For the cliff walking problem, the cells to the south of the bottom row of cells, except for the start and destination cells, form a cliff where, if the agent enters, the episode ends with catastrophic fall into the cliff. Likewise, if the agent tries to leave the left, top, or right boundaries of the grid of cells, it is placed back in the same cell, that is, it is equivalent to taking no action.

For the grid world problem, we do not have a cliff, but we have obstacles inside the grid world. If the agent tries to enter any of these obstacle cells, it is bounced back to the same cell from which it came. In both these problems, the goal is to find the optimum path from the start to the destination.

So, let's dive on in!

Cliff walking with SARSA

We will now learn how to code the aforementioned equations in Python and implement the cliff walking problem with SARSA. First, let's import the `numpy`, `sys`, and `matplotlib` packages in Python. If you have not used these packages in the past, there are several Packt books on these topics to help you come up to speed. Type the following command to install the required packages in a Linux Terminal:

```
sudo apt-get install python-numpy python-scipy python-matplotlib
```

We will now summarize the code involved to solve the grid world problem. In a Terminal, use your favorite editor (for example, gedit, emacs, or vi) to code the following:

```
import numpy as np
import sys
import matplotlib.pyplot as plt
```

We will use a 3 x 12 grid for the cliff walking problem, that is, 3 rows and 12 columns. We also have 4 actions to take at any cell. You can go north, east, south, or west:

```
nrows = 3
ncols = 12
nact = 4
```

We will consider `100000` episodes in total. For exploration, we will use ε-greedy with a value of $\varepsilon = 0.1$. We will consider a constant value for ε, although the interested reader is encouraged to consider variable values for ε as well with its value slowly annealed to zero over the course of the episodes.

The learning rate, α, is chosen as 0.1, and the discount factor $\gamma = 0.95$ is used, which are typical values for this problem:

```
nepisodes = 100000
epsilon = 0.1
alpha = 0.1
gamma = 0.95
```

We will next assign values for the rewards. For any normal action that does not fall into the cliff, the reward is -1; if the agent falls down the cliff, the reward is -100; for reaching the destination, the reward is also -1. Feel free to explore other values for these rewards later, and investigate its effect on the final Q values and path taken from start to destination:

```
reward_normal = -1
reward_cliff = -100
reward_destination = -1
```

The Q values for state-action pairs are initialized to zero. We will use a NumPy array for Q, which is `nrows x ncols x nact`, that is, we have a `nact` number of entries for each cell, and we have `nrows x ncols` total number of cells:

```
Q = np.zeros((nrows,ncols,nact),dtype=np.float)
```

We will define a function to make the agent go to the start location, which has (x, y) co-ordinates of $(x=0, y=nrows)$:

```
def go_to_start():
    # start coordinates
    y = nrows
    x = 0
    return x, y
```

Next, we define a function to take a random action, where we define `a = 0` for moving `top/north`, `a = 1` for moving `right/east`, `a = 2` for moving `bottom/south`, and `a = 4` for moving `left/west`. Specifically, we will use NumPy's `random.randint()` function, as follows:

```
def random_action():
    # a = 0 : top/north
    # a = 1 : right/east
    # a = 2 : bottom/south
    # a = 3 : left/west
    a = np.random.randint(nact)
    return a
```

We will now define the move function, which will take a given (*x*, *y*) location of the agent and the current action, a, and then will perform the action. It will return the new location of the agent after taking the action, (*x1*, *y1*), as well as the state of the agent, which we define as state = 0 for the agent to be OK after taking the action; state = 1 for reaching the destination; and state = 2 for falling into the cliff. If the agent leaves the domain through the left, top, or right, it is sent back to the same grid (equivalent to taking no action):

```
def move(x,y,a):
    # state = 0: OK
    # state = 1: reached destination
    # state = 2: fell into cliff
    state = 0

   if (x == 0 and y == nrows and a == 0):
     # start location
     x1 = x
     y1 = y - 1
     return x1, y1, state
   elif (x == ncols-1 and y == nrows-1 and a == 2):
     # reached destination
     x1 = x
     y1 = y + 1
     state = 1
     return x1, y1, state
   else:
     # inside grid; perform action
     if (a == 0):
       x1 = x
       y1 = y - 1
     elif (a == 1):
       x1 = x + 1
       y1 = y
     elif (a == 2):
       x1 = x
       y1 = y + 1
     elif (a == 3):
       x1 = x - 1
       y1 = y

     # don't allow agent to leave boundary
     if (x1 < 0):
       x1 = 0
     if (x1 > ncols-1):
       x1 = ncols-1
     if (y1 < 0):
       y1 = 0
     if (y1 > nrows-1):
```

```
        state = 2

    return x1, y1, state
```

We will next define the `exploit` function, which will take the (*x, y*) location of the agent and take the greedy action based on the Q values, that is, it will take the a action that has the highest Q value at that (*x, y*) location. We will do this using NumPy's `np.argmax()`. If we are at the start location, we go north (`a = 0`); if we are one step away from the destination, we go south (`a = 2`):

```python
def exploit(x,y,Q):
    # start location
    if (x == 0 and y == nrows):
        a = 0
        return a

    # destination location
    if (x == ncols-1 and y == nrows-1):
        a = 2
        return a

    if (x == ncols-1 and y == nrows):
        print("exploit at destination not possible ")
        sys.exit()

    # interior location
    if (x < 0 or x > ncols-1 or y < 0 or y > nrows-1):
        print("error ", x, y)
        sys.exit()

    a = np.argmax(Q[y,x,:])
    return a
```

Next, we will perform the Bellman update using the following `bellman()` function:

```python
def bellman(x,y,a,reward,Qs1a1,Q):
    if (y == nrows and x == 0):
        # at start location; no Bellman update possible
        return Q

    if (y == nrows and x == ncols-1):
        # at destination location; no Bellman update possible
        return Q

    # perform the Bellman update
    Q[y,x,a] = Q[y,x,a] + alpha*(reward + gamma*Qs1a1 - Q[y,x,a])
    return Q
```

We will then define a function to explore or exploit, depending on a random number less than ε, the parameter we use in the ε-greedy exploration strategy. For this, we will use NumPy's `np.random.uniform()`, which will output a real random number between zero and one:

```
def explore_exploit(x,y,Q):
    # if we end up at the start location, then exploit
    if (x == 0 and y == nrows):
        a = 0
        return a

    # call a uniform random number
    r = np.random.uniform()

    if (r < epsilon):
        # explore
        a = random_action()
    else:
        # exploit
        a = exploit(x,y,Q)
    return a
```

We now have all the functions required for the cliff walking problem. So, we will loop over the episodes and for each episode we start at the starting location, then explore or exploit, then we move the agent one step depending on the action. Here is the Python code for this:

```
for n in range(nepisodes+1):

    # print every 1000 episodes
    if (n % 1000 == 0):
        print("episode #: ", n)

    # start
    x, y = go_to_start()

    # explore or exploit
    a = explore_exploit(x,y,Q)

    while(True):
        # move one step
        x1, y1, state = move(x,y,a)
```

We perform the Bellman update based on the rewards received; note that this is based on the equations presented earlier in this chapter in the theory section. We stop the episode if we reach the destination or fall down the cliff; if not, we continue the exploration or exploitation strategy for one more step, and this goes on and on. The `state` variable in the following code takes the `1` value for reaching the destination, takes the value `2` for falling down the cliff, and is `0` otherwise:

```
# Bellman update
if (state == 1):
    reward = reward_destination
    Qs1a1 = 0.0
    Q = bellman(x,y,a,reward,Qs1a1,Q)
    break
elif (state == 2):
    reward = reward_cliff
    Qs1a1 = 0.0
    Q = bellman(x,y,a,reward,Qs1a1,Q)
    break
elif (state == 0):
    reward = reward_normal
    # Sarsa
    a1 = explore_exploit(x1,y1,Q)
    if (x1 == 0 and y1 == nrows):
        # start location
        Qs1a1 = 0.0
    else:
        Qs1a1 = Q[y1,x1,a1]

    Q = bellman(x,y,a,reward,Qs1a1,Q)
    x = x1
    y = y1
    a = a1
```

The preceding code will complete all the episodes, and we now have the converged values of Q. We will now plot this using `matplotlib` for each of the actions:

```
for i in range(nact):
    plt.subplot(nact,1,i+1)
    plt.imshow(Q[:,:,i])
    plt.axis('off')
    plt.colorbar()
    if (i == 0):
        plt.title('Q-north')
    elif (i == 1):
        plt.title('Q-east')
    elif (i == 2):
        plt.title('Q-south')
```

```
elif (i == 3):
  plt.title('Q-west')
plt.savefig('Q_sarsa.png')
plt.clf()
plt.close()
```

Finally, we will do a path planning using the preceding converged Q values. That is, we will plot the exact path the agent will take from start to finish using the final converged Q values. For this, we will create a variable called path, and store values for it tracing the path. We will then plot it using matplotlib as follows:

```
path = np.zeros((nrows,ncols,nact),dtype=np.float)
x, y = go_to_start()
while(True):
 a = exploit(x,y,Q)
 print(x,y,a)
 x1, y1, state = move(x,y,a)
 if (state == 1 or state == 2):
 print("breaking ", state)
 break
 elif (state == 0):
 x = x1
 y = y1
 if (x >= 0 and x <= ncols-1 and y >= 0 and y <= nrows-1):
 path[y,x] = 100.0
plt.imshow(path)
plt.savefig('path_sarsa.png')
```

That's it. We have completed the coding required for the cliff walking problem with SARSA. We will now view the results. In the following screenshot, we present the Q values for each of the actions (going north, east, south, or west) at each of the locations in the grid. As per the legend, yellow represents high Q values and violet represents low Q values.

SARSA clearly tries to avoid the cliff by choosing to not go south when the agent is just one step to the north of the cliff, as is evident from the large negative Q values for the south action:

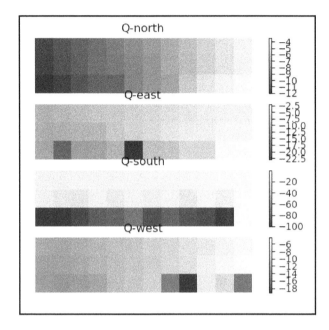

Figure 1: Q values for the cliff walking problem using SARSA

We will next plot the path taken by the agent from start to finish in the following screenshot:

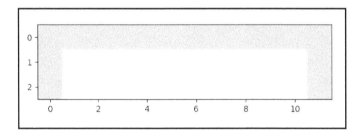

Figure 2: Path taken by the agent using SARSA

The same cliff walking problem will now be investigated using Q-learning.

Cliff walking with Q-learning

We will now repeat the same cliff walking problem, albeit using Q-learning in lieu of SARSA. Most of the code is the same as was used for SARSA, except for a few differences that will be summarized here. Since Q-learning uses a greedy action selection strategy, we will use a function for this as follows to compute the value of the maximum value of Q at a given location. Most of the code is the same as in the previous section, so we will only specify the changes to be made.

Now, let's code cliff walking with Q-learning.

The `max_Q()` function is defined as follows:

```
def max_Q(x,y,Q):
    a = np.argmax(Q[y,x,:])
    return Q[y,x,a]
```

We will compute the Q value at the new state using the `max_Q()` function defined previously:

```
Qs1a1 = max_Q(x1,y1,Q)
```

In addition, the action choosing whether it is exploration or exploitation is done inside the `while` loop as we choose actions greedily when exploiting:

```
# explore or exploit
a = explore_exploit(x,y,Q)
```

That's it for coding Q-learning. We will now apply this to solve the cliff walking problem and present the Q values for each of the actions, and the path traced by the agent to go from start to finish, which are shown in the following screenshots:

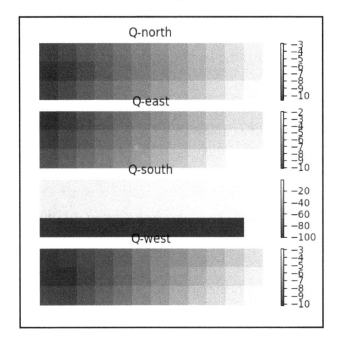

Figure 3: Q values for the cliff walking problem using Q-learning

As evident, the path traced is now different for Q-learning vis-à-vis SARSA. Since Q-learning is a greedy strategy, the agent now takes a path close to the cliff at the bottom of the following screenshot (*Figure 4*), as it is the shortest path:

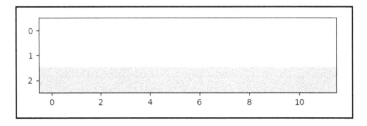

Figure 4: Path traced by the agent for the cliff walking problem using Q-learning

On the other hand, since SARSA is more far-sighted, and so chooses the safe but longer path that is the top row of cells (see *Figure 2*).

Our next problem is the grid world problem, where we must navigate a grid. We will code this in SARSA.

Grid world with SARSA

We will next consider the grid world problem, and we will use SARSA to solve it. We will introduce obstacles in place of a cliff. The goal of the agent is to navigate the grid world from start to destination by avoiding the obstacles. We will store the coordinates of the obstacle cells in the obstacle_cells list, where each entry is the (x, y) coordinate of the obstacle cell.

Here is a summary of the steps involved in this task:

1. Most of the code is the same as previously used, the differences will be summarized here
2. Place obstacles in the grid
3. The move() function has to also look for obstacles in the grid
4. Plot Q values and the path traced by the agent

Here, we will start coding the algorithm in Python:

```python
import numpy as np
import sys
import matplotlib.pyplot as plt


nrows = 3
ncols = 12
nact = 4

nepisodes = 100000
epsilon = 0.1
alpha = 0.1
gamma = 0.95


reward_normal = -1
reward_destination = -1


# obstacles
obstacle_cells = [(4,1), (4,2), (8,0), (8,1)]
```

The move() function will now change, as we have to also look for obstacles. If the agent ends up in one of the obstacle cells it is pushed back to where it came from, as shown in the following code snippet:

```
def move(x,y,a):
  # state = 0: OK
  # state = 1: reached destination
  state = 0

  if (x == 0 and y == nrows and a == 0):
    # start location
    x1 = x
    y1 = y - 1
    return x1, y1, state
  elif (x == ncols-1 and y == nrows-1 and a == 2):
    # reached destination
    x1 = x
    y1 = y + 1
    state = 1
    return x1, y1, state
  else:

    if (a == 0):
      x1 = x
      y1 = y - 1
    elif (a == 1):
      x1 = x + 1
      y1 = y
    elif (a == 2):
      x1 = x
      y1 = y + 1
    elif (a == 3):
      x1 = x - 1
      y1 = y

    if (x1 < 0):
     x1 = 0
    if (x1 > ncols-1):
     x1 = ncols-1
    if (y1 < 0):
     y1 = 0
    if (y1 > nrows-1):
     y1 = nrows-1

    # check for obstacles; reset to original (x,y) if inside obstacle
    for i in range(len(obstacle_cells)):
      if (x1 == obstacle_cells[i][0] and y1 == obstacle_cells[i][1]):
        x1 = x
```

```
        y1 = y
    return x1, y1, state
```

That's it for coding grid world with SARSA. The Q-values and the path taken are shown in following diagrams:

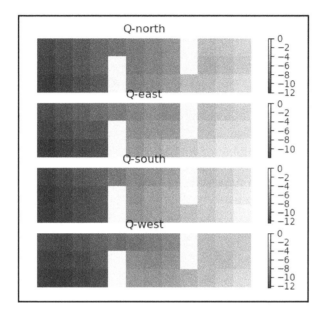

Figure 5: Q-values for each of the actions for the grid world problem using SARSA

As we can see in the following diagram, the agent navigates around the obstacles to reach its destination:

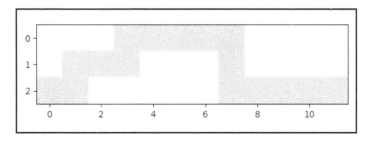

Figure 6: Path traced by the agent for the grid world problem using SARSA

Grid world with Q-learning is not a straightforward problem to attempt, as the greedy strategy used will not enable the agent to avoid repeated actions easily at a given state. Convergence is typically very slow, and so it will be avoided for now.

Summary

In this chapter, we looked at the concept of TD. We also learned about our first two RL algorithms: Q-learning and SARSA. We saw how you can code these two algorithms in Python and use them to solve the cliff walking and grid world problems. These two algorithms give us a good understanding of the basics of RL and how to transition from theory to code. These two algorithms were very popular in the 1990s and early 2000s, before deep RL gained prominence. Despite that, Q-learning and SARSA still find use in the RL community today.

In the next chapter, we will look at the use of deep neural networks in RL that gives rise to deep RL. We will see a variant of Q-learning called **Deep Q-Networks** (**DQNs**) that will use a neural network instead of a tabular state-action value function, which we saw in this chapter. Note that only problems with small number of states and actions are suited to Q-learning and SARSA. When we have a large number of states and/or actions, we encounter what is called as the Curse of Dimensionality, where a tabular approach will be unfeasible due to excessive memory use; in these problems, DQN is better suited, and will be the crux of the next chapter.

Further reading

- *Reinforcement Learning: an Introduction* by *Richard Sutton and Andrew Barto*, 2018

3
Deep Q-Network

Deep Q-Networks (**DQNs**) revolutionized the field of **reinforcement learning** (**RL**). I am sure you have heard of Google DeepMind, which used to be a British company called DeepMind Technologies until Google acquired it in 2014. DeepMind published a paper in 2013 titled *Playing Atari with Deep RL*, where they used **Deep Neural Networks** (**DNNs**) in the context of RL, or DQNs as they are referred to – which is an idea that is seminal to the field. This paper revolutionized the field of deep RL, and the rest is history! Later, in 2015, they published a second paper, titled *Human Level Control Through Deep RL*, in *Nature*, where they had more interesting ideas that further improved the former paper. Together, the two papers led to a Cambrian explosion in the field of deep RL, with several new algorithms that have improved the training of agents using neural networks, and have also pushed the limits of applying deep RL to interesting real-world problems.

In this chapter, we will investigate a DQN and also code it using Python and TensorFlow. This will be our first use of deep neural networks in RL. It will also be our first effort in this book to use deep RL to solve real-world control problems.

In this chapter, the following topics will be covered:

- Learning the theory behind a DQN
- Understanding target networks
- Learning about replay buffer
- Getting introduced to the Atari environment
- Coding a DQN in TensorFlow
- Evaluating the performance of a DQN on Atari Breakout

Technical requirements

Knowledge of the following will help you to better understand the concepts presented in this chapter:

- Python (2 and above)
- NumPy
- TensorFlow (version 1.4 or higher)

Learning the theory behind a DQN

In this section, we will look at the theory behind a DQN, including the math behind it, and learn the use of neural networks to evaluate the `value` function.

Previously, we looked at Q-learning, where *Q(s,a)* was stored and evaluated as a multi-dimensional array, with one entry for each state-action pair. This worked well for grid-world and cliff-walking problems, both of which are low-dimensional in both state and action spaces. So, can we apply this to higher dimensional problems? Well, no, due to the *curse of dimensionality*, which makes it unfeasible to store very large number states and actions. Moreover, in continuous control problems, the actions vary as a real number in a bounded range, although an infinite number of real numbers are possible, which cannot be represented as a tabular *Q* array. This gave rise to function approximations in RL, particularly with the use of DNNs – that is, DQNs. Here, *Q(s,a)* is represented as a DNN that will output the value of *Q*.

The following are the steps that are involved in a DQN:

1. Update the state-action value function using a Bellman equation, where *(s, a)* are the states and actions at a time, *t*, *s'* and *a'* are respectively the states and actions at the subsequent time *t+1*, and *γ* is the discount factor:

$$Q(s,a) = E[r + \gamma \max_{a'} Q(s',a')|s,a]$$

2. We then define a loss function at iteration step *i* to train the Q-network as follows:

$$L_i(\theta_i) = E[(y_i - Q(s, a; \theta_i))^2]$$

The preceding parameters are are the neural network parameters, which are represented as *θ*, hence the Q-value is written as *Q(s, a; θ)*.

3. y_i is the target for iteration *i*, and is given by the following equation:

$$y_i = E[r + \gamma \max_{a'} Q(s', a'; \theta_{i-1}) | s, a]$$

4. We then train the neural network on the DQN by minimizing this loss function *L(θ)* using optimization algorithms, such as gradient descent, RMSprop, and Adam.

We used the least squared loss previously for the DQN loss function, also referred to as the L2 loss. You can also consider other losses, such as the Huber loss, which combines the L1 and L2 losses, with the L2 loss in the vicinity of zero and L1 in regions far away. The Huber loss is less sensitive to outliers than the L2 loss.

We will now look at the use of target networks. This is a very important concept, required to stabilize training.

Understanding target networks

An interesting feature of a DQN is the utilization of a second network during the training procedure, which is referred to as the target network. This second network is used for generating the target-Q values that are used to compute the loss function during training. Why not use just use one network for both estimations, that is, for choosing the action *a* to take, as well as updating the Q-network? The issue is that, at every step of training, the Q-network's values change, and if we use a constantly changing set of values to update our network, then the estimations can easily become unstable – the network can fall into feedback loops between the target and estimated Q-values. In order to mitigate this instability, the target network's weights are fixed – that is, slowly updated to the primary Q-network's values. This leads to training that is far more stable and practical.

We have a second neural network, which we will refer to as the target network. It is identical in architecture to the primary Q-network, although the neural network parameter values are different. Once every N steps, the parameters are copied from the Q-network to the target network. This results in stable training. For example, $N = 10,000$ steps can be used. Another option is to slowly update the weights of the target network (here, θ is the Q-network's weights, and θ^t is the target network's weights):

$$\theta^t \leftarrow \tau\theta + (1 - \tau)\theta^t$$

Here, τ is a small number, say, 0.001. This latter approach of using an exponential moving average is the preferred choice in this book.

Let's now learn about the use of replay buffer in off-policy algorithms.

Learning about replay buffer

We need the tuple (s, a, r, s', done) for updating the DQN, where s and a are respectively the state and actions at time t; s' is the new state at time *t+1*; and done is a Boolean value that is True or False depending on whether the episode is not completed or has ended, also referred to as the terminal value in the literature. This Boolean done or terminal variable is used so that, in the Bellman update, the last terminal state of an episode is properly handled (since we cannot do an *r + γ max Q(s',a')* for the terminal state). One problem in DQNs is that we use contiguous samples of the (s, a, r, s', done) tuple, they are correlated, and so the training can overfit.

To mitigate this issue, a replay buffer is used, where the tuple (s, a, r, s', done) is stored from experience, and a mini-batch of such experiences are randomly sampled from the replay buffer and used for training. This ensures that the samples drawn for each mini-batch are **independent and identically distributed** (**IID**). Usually, a large-size replay buffer is used, say, 500,000 to 1 million samples. At the beginning of the training, the replay buffer is filled to a sufficient number of samples and populated with new experiences. Once the replay buffer is filled to a maximum number of samples, the older samples are discarded one by one. This is because the older samples were generated from an inferior policy, and are not desired for training at a later stage as the agent has advanced in its learning.

In a more recent paper, DeepMind came up with a prioritized replay buffer, where the absolute value of the temporal difference error is used to give importance to a sample in the buffer. Thus, samples with higher errors have a higher priority and so have a bigger chance of being sampled. This prioritized replay buffer results in faster learning than the vanilla replay buffer. However, it is slightly harder to code, as it uses a SumTree data structure, which is a binary tree where the value of every parent node is the sum of the values of its two child nodes. This prioritized experience replay will not be discussed further for now!

 The prioritized experience replay buffer is based on this DeepMind paper: https://arxiv.org/abs/1511.05952

We will now look into the Atari environment. If you like playing video games, you will love this section!

Getting introduced to the Atari environment

The Atari 2600 game suite was originally released in the 1970s, and was a big hit at that time. It involves several games that are played by users using the keyboard to enter actions. These games were a big hit back in the day, and inspired many computer game players of the 1970s and 1980s, but are considered too primitive by today's video game players' standards. However, they are popular today in the RL community as a portal to games that can be trained by RL agents.

Summary of Atari games

Here is a summary of a select few games from Atari (we won't present screenshots of the games for copyright reasons, but will provide links to them).

Pong

Our first example is a ping pong game called Pong, which allows the user to move up or down to hit a ping pong ball to an opponent, which is the computer. The first one to score 21 points is the winner of the game. A screenshot of the Pong game from Atari can be found at https://gym.openai.com/envs/Pong-v0/.

Breakout

In another game, called Breakout, the user must move a paddle to the left or right to hit a ball that then bounces off a set of blocks at the top of the screen. The higher the number of blocks hit, the more points or rewards the player can accrue. There are a total of five lives per game, and if the player misses the ball, it results in the loss of a life. A screenshot of the Breakout game from Atari can be found at https://gym.openai.com/envs/Breakout-v0/.

Space Invaders

If you like shooting space aliens, then Space Invaders is the game for you. In this game, wave after wave of space aliens descend from the top, and the goal is to shoot them using a laser beam, accruing points. The link to this can be found at https://gym.openai.com/envs/SpaceInvaders-v0/.

LunarLander

Or, if you are fascinated by space travel, then LunarLander is about landing a spacecraft (which resembles the Apollo 11 Eagle) on the surface of the moon. For each level, the surface of the landing zone changes and the goal is to guide the spacecraft to land on the lunar surface between two flags. A screenshot of LunarLander from Atari can be found at https://gym.openai.com/envs/LunarLander-v2/.

The Arcade Learning Environment

Over 50 such games exist in Atari. They are now part of the **Arcade Learning Environment (ALE)**, which is an object-oriented framework built on top of Atari. OpenAI's gym is used to invoke Atari games these days so that RL agents can be trained to play these games. For instance, you can import gym in Python and play them as follows.

The reset() function resets the game environment, and render() renders the screenshot of the game:

```
import gym
env = gym.make('SpaceInvaders-v0')
env.reset()
env.render()
```

We will now code a DQN in TensorFlow and Python to train an agent on how to play Atari Breakout.

Coding a DQN in TensorFlow

Here, we will code a DQN using TensorFlow and play Atari Breakout. There are three Python files that we will use:

- `dqn.py`: This file will have the main loop, where we explore the environment and call the update functions
- `model.py`: This file will have the class for the DQN agent, where we will have the neural network and the functions we require to train it
- `funcs.py`: This file will involve some utility functions—for example, to process the image frames, or to populate the replay buffer

Using the model.py file

Let's first code the `model.py` file. The steps involved in this are as follows:

1. **Import the required packages**:

```
import numpy as np
import sys
import os
import random
import tensorflow as tf
```

2. **Choose the bigger or smaller network**: We will use two neural network architectures, one called `bigger` and the other `smaller`. Let's use the `bigger` network for now; the interested user can later change the network to the `smaller` option and compare performance:

```
NET = 'bigger' # 'smaller'
```

3. **Choose the loss function (L2 loss or the Huber loss)**: For the Q-learning `loss` function, we can use either the L2 loss or the Huber loss. Both options will be used in the code. We will choose `huber` for now:

```
LOSS = 'huber' # 'L2'
```

4. **Define neural network weights initialization**: We will then specify a weights initializer for the neural network weights. `tf.variance_scaling_initializer(scale=2)` is used for He initialization. Xavier initialization can also be used, and is provided as a comment. The interested user can compare the performance of both the He and Xavier initializers later:

```
init = tf.variance_scaling_initializer(scale=2) #
tf.contrib.layers.xavier_initializer()
```

5. **Define the QNetwork() class**: We will then define the `QNetwork()` class as follows. It will have an `__init__()` constructor and the `_build_model()`, `predict()`, and `update()` functions. The `__init__` constructor is shown as follows:

```
class QNetwork():
 def __init__(self, scope="QNet", VALID_ACTIONS=[0, 1, 2, 3]):
 self.scope = scope
 self.VALID_ACTIONS = VALID_ACTIONS
 with tf.variable_scope(scope):
 self._build_model()
```

6. **Complete the _build_model() function**: In `_build_model()`, we first define the TensorFlow `tf_X`, `tf_Y`, and `tf_actions` placeholders. Note that the image frames are stored in `uint8` format in the replay buffer to save memory, and so they are normalized by converting them to `float` and then dividing them by `255.0` to put the X input in the 0-1 range:

```
def _build_model(self):
  # input placeholders; input is 4 frames of shape 84x84
  self.tf_X = tf.placeholder(shape=[None, 84, 84, 4],
dtype=tf.uint8, name="X")
  # TD
  self.tf_y = tf.placeholder(shape=[None], dtype=tf.float32,
name="y")
  # action
  self.tf_actions = tf.placeholder(shape=[None], dtype=tf.int32,
name="actions")
  # normalize input
  X = tf.to_float(self.tf_X) / 255.0
  batch_size = tf.shape(self.tf_X)[0]
```

7. **Defining convolutional layers**: As mentioned earlier, we have a `bigger` and a `smaller` neural network option. The `bigger` network has three convolutional layers, followed by a fully connected layer. The `smaller` network only has two convolutional layers, followed by a fully connected layer. We can define convolutional layers in TensorFlow using `tf.contrib.layers.conv2d()`, and fully connected layers using `tf.contrib.layers.fully_connected()`. Note that, after the last convolutional layer, we need to flatten the output before passing it to the fully connected layer, for which we will use `tf.contrib.layers.flatten()`. We use the `winit` object as our weights initializer, which we defined earlier:

```
if (NET == 'bigger'):
  # bigger net
  # 3 conv layers
  conv1 = tf.contrib.layers.conv2d(X, 32, 8, 4, padding='VALID',
activation_fn=tf.nn.relu, weights_initializer=winit)
  conv2 = tf.contrib.layers.conv2d(conv1, 64, 4, 2,
padding='VALID', activation_fn=tf.nn.relu,
weights_initializer=winit)
  conv3 = tf.contrib.layers.conv2d(conv2, 64, 3, 1,
padding='VALID', activation_fn=tf.nn.relu,
weights_initializer=winit)
  # fully connected layers
  flattened = tf.contrib.layers.flatten(conv3)
  fc1 = tf.contrib.layers.fully_connected(flattened, 512,
activation_fn=tf.nn.relu, weights_initializer=winit)

  elif (NET == 'smaller'):

  # smaller net
  # 2 conv layers
  conv1 = tf.contrib.layers.conv2d(X, 16, 8, 4, padding='VALID',
activation_fn=tf.nn.relu, weights_initializer=winit)
  conv2 = tf.contrib.layers.conv2d(conv1, 32, 4, 2,
padding='VALID',activation_fn=tf.nn.relu,
weights_initializer=winit)
  # fully connected layers
  flattened = tf.contrib.layers.flatten(conv2)
  fc1 = tf.contrib.layers.fully_connected(flattened, 256,
activation_fn=tf.nn.relu, weights_initializer=winit)
```

8. **Defining the fully connected layer**: Finally, we have a fully connected layer sized according to the number of actions, which is specified using `len(self.VALID_ACTIONS)`. The output of this last fully connected layer is stored in `self.predictions`, and represents $Q(s,a)$, which we saw in the equations presented earlier, in the *Learning the theory behind a DQN* section. The actions we pass to this function (`self.tf_actions`) have to be converted to one-hot format, for which we use `tf.one_hot()`. Note that `one_hot` is a way to represent the action number as a binary array with zero for all actions, except for one action, for which we store *a* as `1.0`. Then, we multiply the predictions with the one-hot actions using `self.predictions * action_one_hot`, which is summed over using `tf.reduce_sum()`; this is stored in the `self.action_predictions` variable:

```
# Q(s,a)
  self.predictions = tf.contrib.layers.fully_connected(fc1,
len(self.VALID_ACTIONS), activation_fn=None,
weights_initializer=winit)
  action_one_hot = tf.one_hot(self.tf_actions,
tf.shape(self.predictions)[1], 1.0, 0.0, name='action_one_hot')
  self.action_predictions = tf.reduce_sum(self.predictions *
action_one_hot, reduction_indices=1, name='act_pred')
```

9. **Computing loss for training the Q-network**: We compute the loss for training the Q-network, stored in `self.loss`, using either the L2 loss or the Huber loss, which is determined using the `LOSS` variable. For L2 loss, we use the `tf.squared_difference()` function; for the Huber loss, we use `huber_loss()`, which we will soon define. The loss is averaged over many samples, and for this we use the `tf.reduce_mean()` function. Note that we will compute the loss between the `tf_y` placeholder that we defined earlier and the `action_predictions` variable that we obtained in the previous step:

```
if (LOSS == 'L2'):
   # L2 loss
   self.loss = tf.reduce_mean(tf.squared_difference(self.tf_y,
self.action_predictions), name='loss')
elif (LOSS == 'huber'):
   # Huber loss
   self.loss = tf.reduce_mean(huber_loss(self.tf_y-
self.action_predictions), name='loss')
```

10. **Using the optimizer**: We use either the RMSprop or Adam optimizer, and store it in `self.optimizer`. Our learning objective is to minimize `self.loss`, and so we use `self.optimizer.minimize()`. This is stored in `self.train_op`:

```
# optimizer
  #self.optimizer =
tf.train.RMSPropOptimizer(learning_rate=0.00025, momentum=0.95,
epsilon=0.01)
  self.optimizer = tf.train.AdamOptimizer(learning_rate=2e-5)
  self.train_op=
self.optimizer.minimize(self.loss,global_step=tf.contrib.framework.
get_global_step())
```

11. **Define the predict() function for the class**: In the `predict()` function, we run the `self.predictions` function defined earlier using TensorFlow's `sess.run()`, where `sess` is the `tf.Session()` object that is passed to this function. The states are passed as an argument to this function in the `s` variable, which is passed on to the TensorFlow placeholder, `tf_X`:

```
def predict(self, sess, s):
    return sess.run(self.predictions, { self.tf_X: s})
```

12. **Define the update() function for the class**: Finally, in the `update()` function, we call the `train_op` and `loss` objects, and feed the a dictionary to the placeholders involved in performing these operations, which we call `feed_dict`. The states are stored in `s`, the actions in `a`, and the targets in `y`:

```
def update(self, sess, s, a, y):
    feed_dict = { self.tf_X: s, self.tf_y: y, self.tf_actions: a }
    _, loss = sess.run([self.train_op, self.loss], feed_dict)
    return loss
```

13. **Define the huber_loss() function outside the class**: The last thing to complete `model.py` is the definition of the Huber loss function, which is a blend of L1 and L2 losses. Whenever the input is < `1.0`, the L2 loss is used, and the L1 loss otherwise:

```
# huber loss
def huber_loss(x):
 condition = tf.abs(x) < 1.0
 output1 = 0.5 * tf.square(x)
 output2 = tf.abs(x) - 0.5
 return tf.where(condition, output1, output2)
```

Using the funcs.py file

We will next code `funcs.py` by completing the following steps:

1. **Import packages**: First, we import the required packages:

```
import numpy as np
import sys
import tensorflow as tf
```

2. **Complete the ImageProcess() class**: Then, convert the 210 x 160 x 3 RGB image from the Atari emulator to an 84 x 84 grayscale image. For this, we create an `ImageProcess()` class and use TensorFlow utility functions, such as `rgb_to_grayscale()` to convert RGB to grayscale, `crop_to_bounding_box()` to crop the image to the region of interest, `resize_images()` to resize the image to the desired 84 x 84 size, and `squeeze()` to remove a dimension from the input. The `process()` function of the class will carry out the operations by invoking the `sess.run()` function on `self.output`; note that we pass the `state` variable as a dictionary:

```
# convert raw Atari RGB image of size 210x160x3 into 84x84
grayscale image
class ImageProcess():
  def __init__(self):
    with tf.variable_scope("state_processor"):
      self.input_state = tf.placeholder(shape=[210, 160, 3],
dtype=tf.uint8)
      self.output = tf.image.rgb_to_grayscale(self.input_state)
      self.output = tf.image.crop_to_bounding_box(self.output, 34,
0, 160, 160)
      self.output = tf.image.resize_images(self.output, [84, 84],
method=tf.image.ResizeMethod.NEAREST_NEIGHBOR)
      self.output = tf.squeeze(self.output)

  def process(self, sess, state):
    return sess.run(self.output, { self.input_state: state })
```

3. **Copy model parameters from one network to another**: The next step is to write a function called `copy_model_parameters()`, which will take as arguments the `tf.Session()` object `sess`, and two networks (in this case, the Q-network and the target network). Let's call them `qnet1` and `qnet2`. The function will copy the parameter values from `qnet1` to `qnet2`:

```
# copy params from qnet1 to qnet2
def copy_model_parameters(sess, qnet1, qnet2):
```

```
    q1_params = [t for t in tf.trainable_variables() if
t.name.startswith(qnet1.scope)]
    q1_params = sorted(q1_params, key=lambda v: v.name)
    q2_params = [t for t in tf.trainable_variables() if
t.name.startswith(qnet2.scope)]
    q2_params = sorted(q2_params, key=lambda v: v.name)
    update_ops = []
    for q1_v, q2_v in zip(q1_params, q2_params):
        op = q2_v.assign(q1_v)
        update_ops.append(op)
    sess.run(update_ops)
```

4. **Write a function to use ε-greedy strategy to explore or exploit**: We will then write a function called `epsilon_greedy_policy()`, which will either explore or exploit depending on whether a random real number computed using NumPy's `np.random.rand()` is less than `epsilon`, the parameter described earlier for the ε-greedy strategy. For exploration, all actions have equal probabilities and equal one/`(num_actions)`, where `num_actions` is the number of actions (which is four for Breakout). On the other hand, for exploiting, we use Q-network's `predict()` function to obtain Q values and identify which action has the highest Q value with the use of NumPy's `np.argmax()` function. The output of this function is the probability of each of the actions, which, for exploitation, will have all `0` actions except the one action corresponding to the largest Q value, for which the probability is assigned `1.0`:

```
# epsilon-greedy
def epsilon_greedy_policy(qnet, num_actions):
    def policy_fn(sess, observation, epsilon):
        if (np.random.rand() < epsilon):
            # explore: equal probabiities for all actions
            A = np.ones(num_actions, dtype=float) /
float(num_actions)
        else:
            # exploit
            q_values = qnet.predict(sess, np.expand_dims(observation,
0))[0]
            max_Q_action = np.argmax(q_values)
            A = np.zeros(num_actions, dtype=float)
            A[max_Q_action] = 1.0
        return A
    return policy_fn
```

5. **Write a function to populate the replay memory**: Finally, we will write the `populate_replay_mem` function to populate the replay buffer with `replay_memory_init_size` number of samples. First, we reset the environment using `env.reset()`. Then, we process the state obtained from the reset. We need four frames for each state, as the agent otherwise has no way of determining which way the ball or the paddle are moving, their speed and/or acceleration (in the Breakout game; for other games, such as Space Invaders, similar reasoning applies to determine when and where to fire). For the first frame, we stack up four copies. We also compute `delta_epsilon`, which is the amount epsilon is decreased per time step. The replay memory is initialized as an empty list:

```
# populate replay memory
def populate_replay_mem(sess, env, state_processor,
replay_memory_init_size, policy, epsilon_start, epsilon_end,
epsilon_decay_steps, VALID_ACTIONS, Transition):
    state = env.reset()
    state = state_processor.process(sess, state)
    state = np.stack([state] * 4, axis=2)

    delta_epsilon = (epsilon_start -
epsilon_end)/float(epsilon_decay_steps)

    replay_memory = []
```

6. **Computing action probabilities**: Then, we loop over `replay_memory_init_size` four times, decrease epsilon by `delta_epsilon`, and compute the action probabilities, stored in the `action_probs` variable, using `policy()`, which was passed as an argument. The exact action is determined from the `action_probs` variable by sampling using NumPy's `np.random.choice`. Then, `env.render()` renders the environment, and then we pass the action to `env.step()`, which outputs the next state (stored in `next_state`), the reward for the transition, and whether the episode terminated, which is stored in the Boolean `done` variable.

7. **Append to replay buffer**: We then process the next state and append it to the replay memory, the tuple (`state`, `action`, `reward`, `next_state`, `done`). If the episode is done, we reset the environment to a new round of the game, process the image and stack up four times, as done earlier. If the episode is not yet complete, the new state becomes the current state for the next time step, and we proceed this way on and on until the loop finishes:

```
for i in range(replay_memory_init_size):
        epsilon = max(epsilon_start - float(i) * delta_epsilon,
```

```
epsilon_end)
        action_probs = policy(sess, state, epsilon)
        action = np.random.choice(np.arange(len(action_probs)),
p=action_probs)

        env.render()
        next_state, reward, done, _ =
env.step(VALID_ACTIONS[action])

        next_state = state_processor.process(sess, next_state)
        next_state = np.append(state[:,:,1:],
np.expand_dims(next_state, 2), axis=2)
        replay_memory.append(Transition(state, action, reward,
next_state, done))

        if done:
            state = env.reset()
            state = state_processor.process(sess, state)
            state = np.stack([state] * 4, axis=2)
        else:
            state = next_state
    return replay_memory
```

This completes `funcs.py`.

Using the dqn.py file

We will first code the `dqn.py` file. This requires the following steps:

1. **Import the necessary packages**: We will import the required packages as follows:

```
import gym
import itertools
import numpy as np
import os
import random
import sys
import matplotlib.pyplot as plt
import tensorflow as tf
from collections import deque, namedtuple
from model import *
from funcs import *
```

2. **Set the game and choose the valid actions**: We will then set the game. Let's choose the `BreakoutDeterministic-v4` game for now, which is a later version of Breakout v0. This game has four actions, numbered zero to three, and they represent 0: no-operation (`noop`), 1: `fire`, 2: move left, and 3: move right:

```
GAME = "BreakoutDeterministic-v4" # "BreakoutDeterministic-v0"
# Atari Breakout actions: 0 (noop), 1 (fire), 2 (left) and 3
(right)
VALID_ACTIONS = [0, 1, 2, 3]
```

3. **Set the mode (train/test) and the start iterations**: We will then set the mode in the `train_or_test` variable. Let's start with `train` to begin with (you can later set it to `test` to evaluate the model after the training is complete). We will also train from scratch from the 0 iteration:

```
# set parameters for running
train_or_test = 'train' #'test' #'train'
train_from_scratch = True
start_iter = 0
start_episode = 0
epsilon_start = 1.0
```

4. **Create environment**: We will create the environment `env` object, which will create the `GAME` game. `env.action_space.n` will print the number of actions in this game. `env.reset()` will reset the game and output the initial state/observation (note that state and observation in RL parlance are the same and are interchangeable). `observation.shape` will print the shape of the state space:

```
env = gym.envs.make(GAME)
print("Action space size: {}".format(env.action_space.n))
observation = env.reset()
print("Observation space shape: {}".format(observation.shape)
```

5. **Create paths and directories for storing checkpoint files**: We will then create the paths for storing the checkpoint model files and create the directory:

```
# experiment dir
experiment_dir =
os.path.abspath("./experiments/{}".format(env.spec.id))

# create ckpt directory
checkpoint_dir = os.path.join(experiment_dir, "ckpt")
checkpoint_path = os.path.join(checkpoint_dir, "model")
```

```
if not os.path.exists(checkpoint_dir):
  os.makedirs(checkpoint_dir)
```

6. **Define the deep_q_learning() function**: We will next create the
   `deep_q_learning()` function, which will take a long list of arguments that
   involve the TensorFlow session object, the environment, the Q and target
   network objects, and so on. The policy to be followed
   is `epsilon_greedy_policy()`:

```
def deep_q_learning(sess, env, q_net, target_net, state_processor,
num_episodes, train_or_test='train',
train_from_scratch=True,start_iter=0, start_episode=0,
replay_memory_size=250000, replay_memory_init_size=50000,
update_target_net_every=10000, gamma=0.99, epsilon_start=1.0,
epsilon_end=[0.1,0.01], epsilon_decay_steps=[1e6,1e6],
batch_size=32):
    Transition = namedtuple("Transition", ["state", "action",
"reward", "next_state", "done"])

    # policy
    policy = epsilon_greedy_policy(q_net, len(VALID_ACTIONS))
```

7. **Populate the replay memory with experiences encountered with initial random
   actions**: Then, we populate the replay memory with the initial samples:

```
# populate replay memory
 if (train_or_test == 'train'):
   print("populating replay memory")
   replay_memory = populate_replay_mem(sess, env, state_processor,
replay_memory_init_size, policy, epsilon_start, epsilon_end[0],
epsilon_decay_steps[0], VALID_ACTIONS, Transition)
```

8. **Set the epsilon values**: Next, we will set the `epsilon` values. Note that we have
   a double linear function, which will decrease the value of `epsilon`, first from 1
   to 0.1, and then from 0.1 to 0.01, in as many steps, specified in
   `epsilon_decay_steps`:

```
# epsilon start
if (train_or_test == 'train'):
  delta_epsilon1 = (epsilon_start -
epsilon_end[0])/float(epsilon_decay_steps[0])
  delta_epsilon2 = (epsilon_end[0] -
epsilon_end[1])/float(epsilon_decay_steps[1])
  if (train_from_scratch == True):
    epsilon = epsilon_start
  else:
    if (start_iter <= epsilon_decay_steps[0]):
```

```
        epsilon = max(epsilon_start - float(start_iter) *
    delta_epsilon1, epsilon_end[0])
        elif (start_iter > epsilon_decay_steps[0] and start_iter <
    epsilon_decay_steps[0]+epsilon_decay_steps[1]):
        epsilon = max(epsilon_end[0] - float(start_iter) *
    delta_epsilon2, epsilon_end[1])
        else:
        epsilon = epsilon_end[1]
    elif (train_or_test == 'test'):
      epsilon = epsilon_end[1]
```

9. We will then set the total number of time steps:

```
# total number of time steps
total_t = start_iter
```

10. Then, the main loop starts over the episodes from the start to the total number of episodes. We reset the episode, process the first frame, and stack it up 4 times. Then, we will initialize `loss`, `time_steps`, and `episode_rewards` to 0. The total number of lives per episode for Breakout is 5, and so we keep count of it in the `ale_lives` variable. The total number of time steps in this life of the agent is initialized to a large number:

```
for ep in range(start_episode, num_episodes):

        # save ckpt
        saver.save(tf.get_default_session(), checkpoint_path)

        # env reset
        state = env.reset()
        state = state_processor.process(sess, state)
        state = np.stack([state] * 4, axis=2)

        loss = 0.0
        time_steps = 0
        episode_rewards = 0.0
        ale_lives = 5
        info_ale_lives = ale_lives
        steps_in_this_life = 1000000
        num_no_ops_this_life = 0
```

11. **Keeping track of time steps:** We will use an inner `while` loop to keep track of the time steps in a given episode (note: the outer `for` loop is over episodes, and this inner `while` loop is over time steps in the current episode). We will decrease `epsilon` accordingly, depending on whether it is in the 0.1 to 1 range or in the 0.01 to 0.1 range, both of which have different `delta_epsilon` values:

```
while True:
    if (train_or_test == 'train'):
        #epsilon = max(epsilon - delta_epsilon, epsilon_end)
        if (total_t <= epsilon_decay_steps[0]):
            epsilon = max(epsilon - delta_epsilon1, epsilon_end[0])
        elif (total_t >= epsilon_decay_steps[0] and total_t <=
epsilon_decay_steps[0]+epsilon_decay_steps[1]):
            epsilon = epsilon_end[0] - (epsilon_end[0]-
epsilon_end[1]) / float(epsilon_decay_steps[1]) * float(total_t-
epsilon_decay_steps[0])
            epsilon = max(epsilon, epsilon_end[1])
        else:
            epsilon = epsilon_end[1]
```

12. **Updating the target network:** We update the target network if the total number of time steps so far is a multiple of `update_target_net_every`, which is a user-defined parameter. This is accomplished by calling the `copy_model_parameters()` function:

```
# update target net
if total_t % update_target_net_every == 0:
    copy_model_parameters(sess, q_net, target_net)
    print("\n copied params from Q net to target net ")
```

13. At the start of every new life of the agent, we undertake a no-op (corresponding to action probabilities [1, 0, 0, 0]) a random number of times between zero and seven to make the episode different from past episodes, so that the agent gets to see more variations when it explores and learns the environment. This was also done in the original DeepMind paper, and ensures that the agent learns better, since this randomness will ensure that more diversity is experienced. Once we are outside this initial randomness cycle, the actions are taken as per the `policy()` function.

14. Note that we still need to take one fire operation (action probabilities [0, 1, 0, 0]) for one time step at the start of every new life to kick-start the agent. This is a requirement for the ALE framework, without which the frames will freeze. Thus, the life cycle evolves as a 1 fire operation, followed by a random number (between zero and seven) of no-ops, and then the agent uses the `policy` function:

```
time_to_fire = False
if (time_steps == 0 or ale_lives != info_ale_lives):
    # new game or new life
    steps_in_this_life = 0
    num_no_ops_this_life = np.random.randint(low=0,high=7)
    action_probs = [0.0, 1.0, 0.0, 0.0] # fire
    time_to_fire = True
    if (ale_lives != info_ale_lives):
        ale_lives = info_ale_lives
else:
    action_probs = policy(sess, state, epsilon)

steps_in_this_life += 1
if (steps_in_this_life < num_no_ops_this_life and not
time_to_fire):
    # no-op
    action_probs = [1.0, 0.0, 0.0, 0.0] # no-op
```

15. We will then take the action using NumPy's `random.choice`, which will use the `action_probs` probabilities. Then, we render the environment and take one `step`. `info['ale.lives']` will let us know the number of lives remaining for the agent, from which we can ascertain whether the agent lost a life in the current time step. In the DeepMind paper, the rewards were set to +1 or −1 depending on the sign of the reward, so as to be able to compare the different games. This is accomplished using `np.sign(reward)`, which we will not use for now. We will then process `next_state_img` to convert to grayscale of the desired size, which is then appended to the `next_state` vector, which maintains a sequence of four contiguous frames. The rewards obtained are used to increment `episode_rewards`, and we also increment `time_steps`:

```
action = np.random.choice(np.arange(len(action_probs)),
p=action_probs)
env.render()
next_state_img, reward, done, info =
env.step(VALID_ACTIONS[action])
info_ale_lives = int(info['ale.lives'])

# rewards = -1,0,+1 as done in the paper
```

```
#reward = np.sign(reward)

next_state_img = state_processor.process(sess, next_state_img)

# state is of size [84,84,4]; next_state_img is of size[84,84]
#next_state = np.append(state[:,:,1:], np.expand_dims(next_state,
2), axis=2)
next_state = np.zeros((84,84,4),dtype=np.uint8)
next_state[:,:,0] = state[:,:,1]
next_state[:,:,1] = state[:,:,2]
next_state[:,:,2] = state[:,:,3]
next_state[:,:,3] = next_state_img

episode_rewards += reward
time_steps += 1
```

16. **Updating the networks:** Next, if we are in training mode, we update the networks. First, we pop the oldest element in the replay memory if the size is exceeded. Then, we append the recent tuple (state, action, reward, next_state, done) to the replay memory. Note that, if we have lost a life, we treat done = True in the last time step so that the agent learns to avoid losses of lives; without this, done = True is experienced only when the episode ends, that is, when all lives are lost. However, we also want the agent to be self-conscious of the loss of lives.

17. **Sampling a mini-batch from the replay buffer:** We sample a mini-batch from the replay buffer of batch_size. We calculate the Q values of the next state (q_values_next) using the target network, and use it to compute the greedy Q value, which is used to compute the target (y in the equation presented earlier). Once every four time steps, we update the Q-network using q_net.update(); this update frequency is once every four, as it is known to be more stable:

```
if (train_or_test == 'train'):

    # if replay memory is full, pop the first element
    if len(replay_memory) == replay_memory_size:
        replay_memory.pop(0)

    # save transition to replay memory
    # done = True in replay memory for every loss of life
    if (ale_lives == info_ale_lives):
        replay_memory.append(Transition(state, action, reward,
next_state, done))
    else:
        #print('loss of life ')
        replay_memory.append(Transition(state, action, reward,
```

```
next_state, True))

        # sample a minibatch from replay memory
        samples = random.sample(replay_memory, batch_size)
        states_batch, action_batch, reward_batch, next_states_batch,
done_batch = map(np.array, zip(*samples))


        # calculate q values and targets
        q_values_next = target_net.predict(sess, next_states_batch)
        greedy_q = np.amax(q_values_next, axis=1)
        targets_batch = reward_batch +
np.invert(done_batch).astype(np.float32) * gamma * greedy_q

        # update net
        if (total_t % 4 == 0):
            states_batch = np.array(states_batch)
            loss = q_net.update(sess, states_batch, action_batch,
targets_batch)
```

18. We exit the inner `while` loop if `done` = `True`, otherwise, we proceed to the next time step, where the state will now be `new_state` from the previous time step. We can also print on the screen the episode number, time steps for the episode, the total rewards earned in the episode, the current `epsilon`, and the replay buffer size at the end of the episode. These values are also useful for analysis later, and so we store them in a text file called `performance.txt`:

```
if done:
    #print("done: ", done)
    break

state = next_state
total_t += 1

  if (train_or_test == 'train'):
      print('\n Episode: ', ep, '| time steps: ', time_steps, '|
total episode reward: ', episode_rewards, '| total_t: ', total_t,
'| epsilon: ', epsilon, '| replay mem size: ', len(replay_memory))
  elif (train_or_test == 'test'):
      print('\n Episode: ', ep, '| time steps: ', time_steps, '|
total episode reward: ', episode_rewards, '| total_t: ', total_t,
'| epsilon: ', epsilon)

  if (train_or_test == 'train'):
      f = open("experiments/" + str(env.spec.id) +
"/performance.txt", "a+")
      f.write(str(ep) + " " + str(time_steps) + " " +
```

```
str(episode_rewards) + " " + str(total_t) + " " + str(epsilon) +
'\n')
        f.close()
```

19. The next few lines of code will complete `dqn.py`. We reset the TensorFlow graph to begin with using `tf.reset_default_graph()`. Then, we create two instances of the `QNetwork` class, the `q_net` and `target_net` objects. We create a `state_processor` object of the `ImageProcess` class and also create the TensorFlow `saver` object:

```
tf.reset_default_graph()


# Q and target networks
q_net = QNetwork(scope="q",VALID_ACTIONS=VALID_ACTIONS)
target_net = QNetwork(scope="target_q",
VALID_ACTIONS=VALID_ACTIONS)

# state processor
state_processor = ImageProcess()

# tf saver
saver = tf.train.Saver()
```

20. We will now execute the TensorFlow graph by calling `tf.Session()`. If we are starting from scratch in training mode, we have to initialize the variables, which is accomplished by calling `sess.run()` on `tf.global_variables_initializer()`. Otherwise, if we are in test mode, or in training mode but not starting from scratch, we load the latest checkpoint file by calling `saver.restore()`.

21. The `replay_memory_size` parameter is limited by the size of RAM you have at your disposal. The present simulations were undertaken in a 16 GB RAM computer, where `replay_memory_size = 300000` was the limit. If the reader has access to more RAM, a larger value can be used for this parameter. For your information, DeepMind used a replay memory size of 1,000,000. A larger replay memory size is good, as it helps to provide more diversity in training when a mini-batch is sampled:

```
with tf.Session() as sess:

    # load model/ initialize model
    if ((train_or_test == 'train' and train_from_scratch ==
False) or train_or_test == 'test'):
            latest_checkpoint =
tf.train.latest_checkpoint(checkpoint_dir)
```

```
                        print("loading model ckpt
{}...\n".format(latest_checkpoint))
                        saver.restore(sess, latest_checkpoint)
        elif (train_or_test == 'train' and train_from_scratch ==
True):
                        sess.run(tf.global_variables_initializer())



        # run
        deep_q_learning(sess, env, q_net=q_net,
target_net=target_net, state_processor=state_processor,
num_episodes=25000,
train_or_test=train_or_test,train_from_scratch=train_from_scratch,
start_iter=start_iter, start_episode=start_episode,
replay_memory_size=300000, replay_memory_init_size=5000,
update_target_net_every=10000, gamma=0.99,
epsilon_start=epsilon_start, epsilon_end=[0.1,0.01],
epsilon_decay_steps=[1e6,1e6], batch_size=32)
```

That's it—this completes `dqn.py`.

We will now evaluate the performance of the DQN on Atari Breakout.

Evaluating the performance of the DQN on Atari Breakout

Here, we will plot the performance of our DQN algorithm on Breakout using the `performance.txt` file that we wrote in the code previously. We will use `matplotlib` to plot two graphs as follows:

1. Number of time steps per episode versus episode number
2. Total episode reward versus time step number

The steps involved in this are as follows:

1. **Plot number of time steps versus episode number for Atari Breakout using DQN**: First, we plot the number of time steps the agent lasted per episode of training in the following diagram. As we can see, after about 10,000 episodes, the agent has learned to survive for a peak of 2,000 time steps per episode (blue curve). We also plot the exponentially weighted moving average with the degree of weighing, $\alpha = 0.02$, in orange. The average number of time steps lasted is approximately 1,400 per episode at the end of the training:

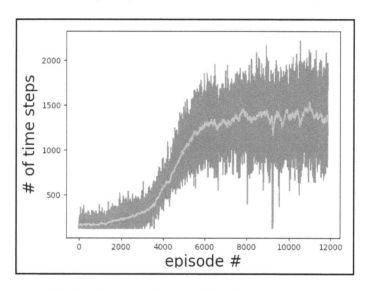

Figure 1: Number of time steps lasted per episode for Atari Breakout using the DQN

2. **Plot episode reward versus time step number**: In the following graph, we plot the total episode reward versus time time step for Atari Breakout using the DQN algorithm. As we can see, the peak episode rewards are close to 400 (blue curve), and the exponentially weighted moving average is approximately 160 to 180 toward the end of the training. We used a replay memory size of 300,000, which is fairly small by modern standards, due to RAM limitations. If a bigger replay memory size was used, a higher average episode reward could be obtained. This is left for experimentation by the reader:

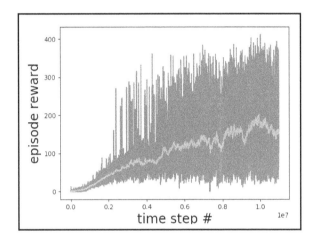

Figure 2: Total episode reward versus time step number for Atari Breakout using the DQN

This concludes this chapter on DQN.

Summary

In this chapter, we looked at our very first deep RL algorithm, DQN, which is probably the most popular RL algorithm in use today. We learned the theory behind a DQN, and also looked at the concept and use of target networks to stabilize training. We were also introduced to the Atari environment, which is the most popular environment suite for RL. In fact, many of the RL papers published today apply their algorithms to games from the Atari suite and report their episodic rewards, comparing them with corresponding values reported by other researchers who use other algorithms. So, the Atari environment is a natural suite of games to train RL agents and compare them to ascertain the robustness of algorithms. We also looked at the use of a replay buffer, and learned why it is used in off-policy algorithms.

This chapter has laid the foundation for us to delve deeper into deep RL (no pun intended!). In the next chapter, we will look at other DQN extensions, such as DDQN, dueling network architectures, and rainbow networks.

Questions

1. Why is a replay buffer used in a DQN?
2. Why do we use target networks?
3. Why do we stack four frames into one state? Will one frame alone suffice to represent one state?
4. Why is the Huber loss sometimes preferred over L2 loss?
5. We converted the RGB input image into grayscale. Can we instead use the RGB image as input to the network? What are the pros and cons of using RGB images instead of grayscale?

Further reading

- *Playing Atari with Deep Reinforcement Learning, by Volodymyr Mnih, Koray Kavukcuoglu, David Silver, Alex Graves, Ioannis Antonoglou, Daan Wierstra, and Martin Riedmiller, arXiv:1312.5602:* https://arxiv.org/abs/1312.5602
- *Human-level control through deep reinforcement learning by Volodymyr Mnih, Koray Kavukcuoglu, David Silver, Andrei A. Rusu, Joel Veness, Marc G. Bellemare, Alex Graves, Martin Riedmiller, Andreas K. Fidjeland, Georg Ostrovski, Stig Petersen, Charles Beattie, Amir Sadik, Ioannis Antonoglou, Helen King, Dharshan Kumaran, Daan Wierstra, Shane Legg, and Demis Hassabis, Nature, 2015:* https://www.nature.com/articles/nature14236

4
Double DQN, Dueling Architectures, and Rainbow

We discussed the **Deep Q-Network** (**DQN**) algorithm in the previous chapter, coded it in Python and TensorFlow, and trained it to play Atari Breakout. In DQN, the same Q-network was used to select and evaluate an action. This, unfortunately, is known to overestimate the Q values, which results in over-optimistic estimates for the values. To mitigate this, DeepMind released another paper where it proposed the decoupling of the action selection and action evaluation. This is the crux of the **Double DQN** (**DDQN**) architectures, which we will investigate in this chapter.

Even later, DeepMind released another paper where they proposed the Q-network architecture with two output values, one representing the value, *V(s)*, and the other the advantage of taking an action at the given state, *A(s,a)*. DeepMind then combined these two to compute the *Q(s,a)* action-value, instead of directly determining it as done in DQN and DDQN. These Q-network architectures are referred to as the **dueling** network architectures, as the neural network now has dual output values, *V(s)* and *A(s,a)*, which are later combined to obtain *Q(s,a)*. We will also see these dueling networks in this chapter.

Another extension we will also consider in this chapter are **Rainbow networks**, which are a blend of several different ideas fused into one algorithm.

The topics that will be covered in this chapter are the following:

- Learning the theory behind DDQN
- Coding DDQN and training it to play Atari Breakout
- Evaluating the performance of DDQN on Atari Breakout
- Understanding dueling network architectures

- Coding dueling network architecture and training it to play Atari Breakout
- Evaluating the performance of dueling architectures on Atari Breakout
- Understanding Rainbow networks
- Running a Rainbow network on Dopamine

Technical requirements

To successfully complete this chapter, knowledge of the following will help significantly:

- Python (2 or 3)
- NumPy
- Matplotlib
- TensorFlow (version 1.4 or higher)
- Dopamine (we will discuss this in more detail later)

Understanding Double DQN

DDQN is an extension to DQN, where we use the target network in the Bellman update. Specifically, in DDQN, we evaluate the target network's Q function using the action that would be greedy maximization of the primary network's Q function. First, we will use the vanilla DQN target for the Bellman equation update step, then, we will extend to DDQN for the same Bellman equation update step; this is the crux of the DDQN algorithm. We will then code DDQN in TensorFlow to play Atari Breakout. Finally, we will compare and contrast the two algorithms: DQN and DDQN.

Updating the Bellman equation

In vanilla DQN, the target for the Bellman update is this:

$$y_t^{DQN} = r_{t+1} + \gamma \max_a Q(s', a; \theta_t)$$

θ_t represents the model parameters of the target network. This is known to over-predict Q, and so the change made in DDQN is to replace this target value, y_t, with this:

$$y_t^{DDQN} = r_{t+1} + \gamma Q(s', \arg\max_a Q(s', a; \theta); \theta_t)$$

We must distinguish between the Q-network parameters, θ, and the target network model parameters, θ_t.

Coding DDQN and training to play Atari Breakout

We will now code DDQN in TensorFlow to play Atari Breakout. As before, we have three Python files:

- funcs.py
- model.py
- ddqn.py

funcs.py and model.py are the same as used before for DQN in Chapter 3, *Deep Q-Network (DQN)*. The ddqn.py file is the only code where we need to make changes to implement DDQN. We will use the same dqn.py file from the previous chapter and make changes to it to code DDQN. So, let's first copy the dqn.py file from before and rename it ddqn.py.

We will summarize the changes we will make to ddqn.py, which are actually quite minimal. We will still not delete the DQN-related lines of code in the file, and instead, use if loops to choose between the two algorithms. This helps to use one code for both algorithms, which is a better way to code.

First, we create a variable called ALGO, which will store one of two strings: DQN or DDQN, which is where we specify which of the two algorithms to use:

```
ALGO = "DDQN" #"DQN" # DDQN
```

Then, in the lines of code where we evaluate the targets for the mini-batch, we use if loops to decide whether the algorithm to use is DQN or DDQN and accordingly compute the targets as follows. Note that, in DQN, the greedy_q variable stores the Q value corresponding to the greedy action taking, that is, the largest Q value in the target network, which is computed using np.amax() and then used to compute the target variable, targets_batch.

In DDQN, on the other hand, we compute the action corresponding to the maximum Q in the primary Q-network, which we store in `greedy_q` and evaluate using `np.argmax()`. Then, we use `greedy_q` (which represents an action now) in the target network Q values. Note that, for Terminal time steps, that is, `done = True`, we should not consider the next state and likewise, for non-Terminal steps, `done = False`, and here we consider the next step. This is easily accomplished using `np.invert().astype(np.float32)` on `done_batch`. The following lines of code show DDQN:

```
# calculate q values and targets

if (ALGO == 'DQN'):

    q_values_next = target_net.predict(sess, next_states_batch)
    greedy_q = np.amax(q_values_next, axis=1)
    targets_batch = reward_batch + np.invert(done_batch).astype(np.float32)
* gamma * greedy_q

elif (ALGO == 'DDQN'):
    q_values_next = q_net.predict(sess, next_states_batch)
    greedy_q = np.argmax(q_values_next, axis=1)
    q_values_next_target = target_net.predict(sess, next_states_batch)
    targets_batch = reward_batch + np.invert(done_batch).astype(np.float32)
* gamma * q_values_next_target[np.arange(batch_size), greedy_q]
```

That's it for `ddqn.py`. We will now evaluate it on Atari Breakout.

Evaluating the performance of DDQN on Atari Breakout

We will now evaluate the performance of DDQN on Atari Breakout. Here, we will plot the performance of our DDQN algorithm on Atari Breakout using the `performance.txt` file that we wrote in the code. We will use `matplotlib` to plot two graphs as explained in the following.

In the following screenshot, we present the number of time steps per episode on Atari Breakout using DDQN and its exponentially weighted moving average. As evident, the peak number of time steps is ~2,000 for many episodes toward the end of the training, with one episode where it exceeded even 3,000 time steps! The moving average is approximately 1,500 time steps toward the end of the training:

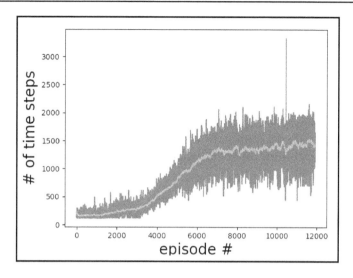

Figure 1: Number of time steps per episode for Atari Breakout using DDQN

In the following screenshot, we show the total rewards received per episode versus the time number of the global time step. The peak episode reward is over 350, with the moving average near 150. Interestingly, the moving average (in orange) is still increasing toward the end, which means you can run the training even longer to see further gains. This is left to the interested reader:

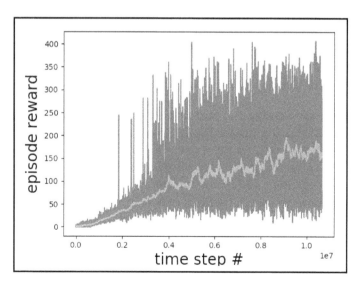

Figure 2: Total episode reward versus time step for Atari Breakout using DDQN

Note that, due to RAM constraints (16 GB), we used a replay buffer size of 300,000 only. If the user has access to more RAM power, a bigger replay buffer size can be used—for example, 500,000 to 1,000,000, which can result in even better scores.

As we can see, the DDQN agent is learning to play Atari Breakout well. The moving average of the episode rewards is constantly going up, which means you can train longer to obtain even higher rewards. This upward trend in the episode reward demonstrates the efficacy of the DDQN algorithm for such problems.

Understanding dueling network architectures

We will now understand the use of dueling network architectures. In DQN and DDQN, and other DQN variants in the literature, the focus was primarily on algorithms, that is, how to efficiently and stably update the value function neural networks. While this is crucial for developing robust RL algorithms, a parallel but complementary direction to advance the field is to also innovate and develop novel neural network architectures that are well suited for model-free RL. This is precisely the concept behind dueling network architectures, another contribution from DeepMind.

The steps involved in dueling architectures are as follows:

1. Dueling network architecture figure; compare with standard DQN
2. Computing $Q(s,a)$
3. Subtracting the average of the advantage from the `advantage` function

As we saw in the previous chapter, the output of the Q-network in DQN is $Q(s,a)$, the action-value function. In dueling networks, the Q-network instead has two output values: the `state value` function, $V(s)$, and the `advantage` function, $A(s,a)$. You can then combine them to compute the `state-action value` function, $Q(s,a)$. This has the advantage that the network need not learn the `value` function for every action at every state. This is particularly useful in states where the actions do not affect the environment.

For instance, if the agent is a car driving on a straight road with no traffic, no action is necessary and so *V(s)* alone will suffice in these states. On the other hand, if the road suddenly curves or other cars come into the vicinity of the agent, then the agent needs to take actions and so, in these states, the `advantage` function comes into play to find the incremental returns a given action can provide over the `state value` function. This is the intuition behind separating the estimation of *V(s)* and *A(s,a)* in the same network by using two different branches, and later combining them.

Refer to the following diagram for a schematic showing a comparison of the standard DQN network and the dueling network architectures:

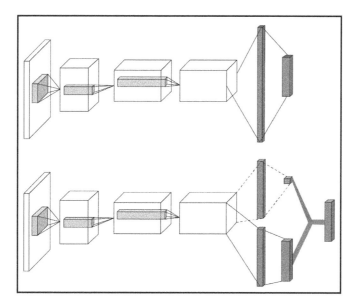

Figure 3: Schematic of the standard DQN network (top) and the dueling network architecture (bottom)

You can compute the `action-value` function *Q(s,a)* as follows:

$$Q(s, a) = V(s) + A(s, a)$$

However, this is not unique in that you can have an amount, *δ*, over-predicted in *V(s)* and the same amount, *δ*, under-predicted in *A(s,a)*. This makes the neural network predictions unidentifiable. To circumvent this problem, the authors of the dueling network paper recommend the following way to combine *V(s)* and *A(s,a)*:

$$Q(s, a; \theta, \alpha, \beta) = V(s; \theta, \beta) + [A(s, a; \theta, \alpha) - \frac{1}{|A|} \sum_{a'} A(s, a'; \theta, \alpha)]$$

$|A|$ represents the number of actions and θ is the neural network parameters that are shared between the $V(s)$ and $A(s,a)$ streams; in addition, α and β are used to denote the neural network parameters in the two different streams, that is, in the $A(s,a)$ and $V(s)$ streams, respectively. Essentially, in the preceding equation, we subtract the average `advantage` function from the `advantage` function and sum it to the `state value` function to obtain $Q(s,a)$.

 This is the link to the dueling network architectures paper: `https://arxiv.org/abs/1511.06581`.

Coding dueling network architecture and training it to play Atari Breakout

We will now code the dueling network architecture and train it to learn to play Atari Breakout. For the dueling network architecture, we require the following codes:

- `model.py`
- `funcs.py`
- `dueling.py`

We will use `funcs.py`, which was used earlier for DDQN, so we reuse it. The `dueling.py` code is also identical to `ddqn.py` (which was used earlier, so we just rename and reuse it). The only changes to be made are in `model.py`. We copy the same `model.py` file from DDQN and summarize here the changes to be made for the dueling network architecture. The steps involved are the following.

We first create a Boolean variable called `DUELING` in `model.py` and assign it to `True` if using dueling network architecture; otherwise, it is assigned to `False`:

```
DUELING = True # False
```

We will write the code with an `if` loop so that the `DUELING` variable, if `False`, will use the earlier code we used in DDQN, and if `True`, we will use the dueling network. We will use the `flattened` object that is the flattened version of the output of the convolutional layers to create two sub-neural network streams. We send `flattened` separately into two different fully connected layers with `512` neurons, using the `relu` activation function and the `winit` weights initializer defined earlier; the output values of these fully connected layers are called `valuestream` and `advantagestream`, respectively:

```
if (not DUELING):

    # Q(s,a)
    self.predictions = tf.contrib.layers.fully_connected(fc1,
len(self.VALID_ACTIONS), activation_fn=None, weights_initializer=winit)

 else:

    # Deuling network
    # branch out into two streams using flattened (i.e., ignore fc1 for
Dueling DQN)

    valuestream = tf.contrib.layers.fully_connected(flattened, 512,
activation_fn=tf.nn.relu, weights_initializer=winit)
    advantagestream = tf.contrib.layers.fully_connected(flattened, 512,
activation_fn=tf.nn.relu, weights_initializer=winit)
```

Combining V and A to obtain Q

The `advantagestream` object is passed into a fully connected layer with a number of neurons equal to the number of actions, that is, `len(self.VALID_ACTIONS)`. Likewise, the `valuestream` object is passed into a fully connected layer with one neuron. Note that we do not use an activation function for computing the `advantage` and `state value` functions, as they can be positive or negative (`relu` will set all negative values to zero!). Finally, we combine the advantage and value streams using `tf.subtract()` to subtract the advantage and the mean of the `advantage` function. The mean is computed using `tf.reduce_mean()` on the `advantage` function:

```
# A(s,a)
self.advantage = tf.contrib.layers.fully_connected(advantagestream,
len(self.VALID_ACTIONS), activation_fn=None, weights_initializer=winit)

# V(s)
self.value = tf.contrib.layers.fully_connected(valuestream, 1,
activation_fn=None, weights_initializer=winit)
```

```
# Q(s,a) = V(s) + (A(s,a) - 1/|A| * sum A(s,a'))
self.predictions = self.value + tf.subtract(self.advantage,
tf.reduce_mean(self.advantage, axis=1, keep_dims=True))
```

That's it for coding dueling network architectures. We will train an agent with the dueling network architecture and evaluate its performance on Atari Breakout. Note that we can use the dueling architecture in conjunction with either DQN or DDQN. That is to say that we only changed the neural network architecture, not the actual Bellman update, and so the dueling architecture works with both DQN and DDQN.

Evaluating the performance of dueling architectures on Atari Breakout

We will now evaluate the performance of dueling architectures on Atari Breakout. Here, we will plot the performance of our dueling network architecture with DDQN on Atari Breakout using the `performance.txt` file that we wrote during the training of the agent. We will use `matplotlib` to plot two graphs as explained in the following.

In the following screenshot, we present the number of time steps per episode on Atari Breakout using DDQN (in blue) and its exponentially weighted moving average (in orange). As evident, the peak number of time steps is ~2,000 for many episodes toward the end of the training, with a few episodes even exceeding 4,000 time steps! The moving average is approximately 1,500 time steps toward the end of the training:

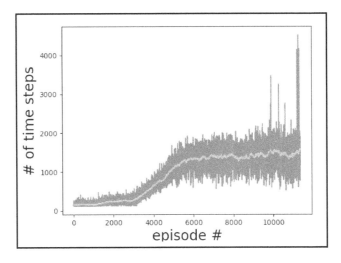

Figure 4: Number of time steps per episode on Atari Breakout using dueling network architecture and DDQN

In the following screenshot, we show the total rewards received per episode versus the time number of the global time step. The peak episode reward is over 400, with the moving average near 220. We also note that the moving average (in orange) is still increasing toward the end, which means you can run the training even longer to obtain further gains. Overall, the average rewards are higher with the dueling network architecture vis-a-vis the non-dueling counterparts, and so it is strongly recommended to use these dueling architectures:

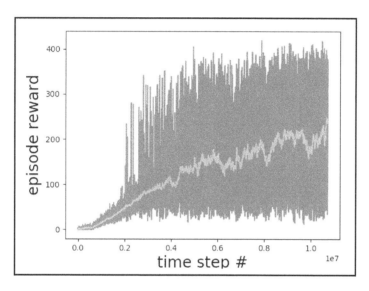

Figure 5: Total episode reward received versus global time step number for Atari Breakout using dueling network architecture and DDQN

 Note that, due to RAM constraints (16 GB), we used a replay buffer size of only 300,000. If the user has access to more RAM power, a bigger replay buffer size can be used—for example, 500,000 to 1,000,000, which can result in even better scores.

Understanding Rainbow networks

We will now move on to **Rainbow networks**, which is a confluence of several different DQN improvements. Since the original DQN paper, several different improvements were proposed with notable success. This motivated DeepMind to combine several different improvements into an integrated agent, which they refer to as the **Rainbow DQN**. Specifically, six different DQN improvements are combined into one integrated Rainbow DQN agent. These six improvements are summarized as follows:

- DDQN
- Dueling network architecture
- Prioritized experience replay
- Multi-step learning
- Distributional RL
- Noisy nets

DQN improvements

We have already seen DDQN and dueling network architectures and have coded them in TensorFlow. The rest of the improvements are described in the following sections.

Prioritized experience replay

We used a replay buffer where all of the samples have an equal probability of being sampled. This, however, is not very efficient, as some samples are more important than others. This is the motivation behind prioritized experience replay, where samples that have a higher **Temporal Difference** (**TD**) error are sampled with a higher probability than others. The first time a sample is added to the replay buffer, it is set a maximum priority value so as to ensure that all samples in the buffer are sampled at least once. Thereafter, the TD error is used to determine the probability of the experience to be sampled, which we compute as this:

$$\text{prob.} \propto |r + \gamma \max_{a'} Q(s', a'; \theta^t) - Q(s, a; \theta)|^\omega$$

Whereas the previous r is the reward, θ is the primary Q-network model parameters, and θ^t is the target network parameters. ω is a positive hyper-parameter that determines the shape of the distribution.

Multi-step learning

In Q-learning, we accumulate a single reward and use the greedy action at the next step. Alternatively, you can also use multi-step targets and compute an *n*-step return from a single state:

$$r_t^{(n)} = \sum_{k=0}^{n-1} \gamma^k r_{t+k+1}$$

Then, the *n*-step return, $r_t^{(n)}$, is used in the Bellman update and is known to lead to faster learning.

Distributional RL

In **distributional RL**, we learn to approximate the distribution of returns instead of the expected return. This is mathematically complicated, is beyond the scope of this book, and is not discussed further.

Noisy nets

In some games (such as Montezuma's revenge), ε-greedy does not work well, as many actions need to be executed before the first reward is received. Under this setting, the use of a noisy linear layer that combined a deterministic and a noisy stream is recommended, shown as follows:

$$y = (b + Wx) + (b^{noisy} \odot \epsilon^b + (W^{noisy} \odot \epsilon^W)x)$$

Here, x is the input, y is the output, and b and W are the biases and weights in the deterministic stream; b^{noisy} and W^{noisy} are the biases and weights, respectively, in the noisy stream; and ϵ^b and ϵ^W are random variables and are applied as element-wise product to the biases and weights, respectively, in the noisy stream. The network may choose to ignore the noisy stream in some regions of the state space and may use them otherwise, as required. This allows for a state-determined exploration strategy.

We will not be coding the full Rainbow DQN, as it is exhaustive. Instead, we will use an open source framework called Dopamine to train a Rainbow DQN agent, which will be discussed in the next section.

Running a Rainbow network on Dopamine

In 2018, some engineers at Google released an open source, lightweight, TensorFlow-based framework for training RL agents, called **Dopamine**. Dopamine, as you may already know, is the name of an organic chemical that plays an important role in the brain. We will use Dopamine to run Rainbow.

The Dopamine framework is based on four design principles:

- Easy experimentation
- Flexible development
- Compact and reliable
- Reproducible

To download Dopamine from GitHub, type the following command in a Terminal:

```
git clone https://github.com/google/dopamine.git
```

We can test whether Dopamine was successfully installed by typing the following commands into a Terminal:

```
cd dopamine
export PYTHONPATH=${PYTHONPATH}:.
python tests/atari_init_test.py
```

The output of this will look something like the following:

```
2018-10-27 23:08:17.810679: I
tensorflow/core/platform/cpu_feature_guard.cc:141] Your CPU supports
instructions that this TensorFlow binary was not compiled to use: SSE4.1
SSE4.2 AVX AVX2 FMA
2018-10-27 23:08:18.079916: I
tensorflow/stream_executor/cuda/cuda_gpu_executor.cc:897] successful NUMA
node read from SysFS had negative value (-1), but there must be at least
one NUMA node, so returning NUMA node zero
2018-10-27 23:08:18.080741: I
tensorflow/core/common_runtime/gpu/gpu_device.cc:1392] Found device 0 with
properties:
name: GeForce GTX 1060 with Max-Q Design major: 6 minor: 1
memoryClockRate(GHz): 1.48
pciBusID: 0000:01:00.0
totalMemory: 5.93GiB freeMemory: 5.54GiB
2018-10-27 23:08:18.080783: I
tensorflow/core/common_runtime/gpu/gpu_device.cc:1471] Adding visible gpu
devices: 0
2018-10-27 23:08:24.476173: I
```

```
tensorflow/core/common_runtime/gpu/gpu_device.cc:952] Device interconnect
StreamExecutor with strength 1 edge matrix:
2018-10-27 23:08:24.476247: I
tensorflow/core/common_runtime/gpu/gpu_device.cc:958] 0
2018-10-27 23:08:24.476273: I
tensorflow/core/common_runtime/gpu/gpu_device.cc:971] 0: N
2018-10-27 23:08:24.476881: I
tensorflow/core/common_runtime/gpu/gpu_device.cc:1084] Created TensorFlow
device (/job:localhost/replica:0/task:0/device:GPU:0 with 5316 MB memory)
-> physical GPU (device: 0, name: GeForce GTX 1060 with Max-Q Design, pci
bus id: 0000:01:00.0, compute capability: 6.1)
.
.
.
Ran 2 tests in 8.475s

OK
```

You should see `OK` at the end to confirm that everything went well with the download.

Rainbow using Dopamine

To run Rainbow DQN, type the following command into a Terminal:

```
python -um dopamine.atari.train --agent_name=rainbow --
base_dir=/tmp/dopamine --
gin_files='dopamine/agents/rainbow/configs/rainbow.gin'
```

That's it. Dopamine will start training Rainbow DQN and print out training statistics on the screen, as well as save checkpoint files. The configuration file is stored in the following path:

dopamine/dopamine/agents/rainbow/configs/rainbow.gin

It looks like the following code. `game_name` is set to `Pong` as default; feel free to try other Atari games. The number of agent steps for training is set in `training_steps`, and for evaluation in `evaluation_steps`. In addition, it introduces stochasticity to the training by using the concept of sticky actions, where the most recent action is repeated multiple times with a probability of 0.25. That is, if a uniform random number (computed using NumPy's `np.random.rand()`) is < 0.25, the most recent action is repeated; otherwise, a new action is taken from the policy.

The sticky action is a new method of introducing stochasticity to the learning:

```
# Hyperparameters follow Hessel et al. (2018), except for sticky_actions,
# which was False (not using sticky actions) in the original paper.
import dopamine.agents.rainbow.rainbow_agent
import dopamine.atari.run_experiment
import dopamine.replay_memory.prioritized_replay_buffer
import gin.tf.external_configurables

RainbowAgent.num_atoms = 51
RainbowAgent.vmax = 10.
RainbowAgent.gamma = 0.99
RainbowAgent.update_horizon = 3
RainbowAgent.min_replay_history = 20000 # agent steps
RainbowAgent.update_period = 4
RainbowAgent.target_update_period = 8000 # agent steps
RainbowAgent.epsilon_train = 0.01
RainbowAgent.epsilon_eval = 0.001
RainbowAgent.epsilon_decay_period = 250000 # agent steps
RainbowAgent.replay_scheme = 'prioritized'
RainbowAgent.tf_device = '/gpu:0' # use '/cpu:*' for non-GPU version
RainbowAgent.optimizer = @tf.train.AdamOptimizer()

# Note these parameters are different from C51's.
tf.train.AdamOptimizer.learning_rate = 0.0000625
tf.train.AdamOptimizer.epsilon = 0.00015

Runner.game_name = 'Pong'
# Sticky actions with probability 0.25, as suggested by (Machado et al.,
2017).
Runner.sticky_actions = True
Runner.num_iterations = 200
Runner.training_steps = 250000 # agent steps
Runner.evaluation_steps = 125000 # agent steps
Runner.max_steps_per_episode = 27000 # agent steps

WrappedPrioritizedReplayBuffer.replay_capacity = 1000000
WrappedPrioritizedReplayBuffer.batch_size = 32
```

Feel free to experiment with the hyperparameters and see how the learning is affected. This is a very nice way to ascertain the sensitivity of the different hyperparameters on the learning of the RL agent.

Summary

In this chapter, we were introduced to DDQN, dueling network architectures, and the Rainbow DQN. We extended our previous DQN code to DDQN and dueling architectures and tried it out on Atari Breakout. We can clearly see that the average episode rewards are higher with these improvements, and so these improvements are a natural choice to use. Next, we also saw Google's Dopamine and used it to train a Rainbow DQN agent. Dopamine has several other RL algorithms, and the user is encouraged to dig deeper and try out these other RL algorithms as well.

This chapter was a good deep dive into the DQN variants, and we really covered a lot of mileage as far as coding of RL algorithms is involved. In the next chapter, we will learn about our next RL algorithm called **Deep Deterministic Policy Gradient** (**DDPG**), which is our first Actor-Critic RL algorithm and our first continuous action space RL algorithm.

Questions

1. Why does DDQN perform better than DQN?
2. How does the dueling network architecture help in the training?
3. Why does prioritized experience replay speed up the training?
4. How do sticky actions help in the training?

Further reading

- The DDQN paper, *Deep Reinforcement Learning with Double Q-learning*, by Hado van Hasselt, Arthur Guez, and David Silver can be obtained from the following link, and the interested reader is recommended to read it: `https://arxiv.org/abs/1509.06461`
- *Rainbow: Combining Improvements in Deep Reinforcement Learning*, Matteo Hessel, Joseph Modayil, Hado van Hasselt, Tom Schaul, Georg Ostrovski, Will Dabney, Dan Horgan, Bilal Piot, Mohammad Azar, and David Silver, arXiv:1710.02298 (the Rainbow DQN): `https://arxiv.org/abs/1710.02298`

- *Prioritized Experience Replay*, Tom Schaul, John Quan, Ioannis Antonoglou, David Silver, arXiv:1511.05952: `https://arxiv.org/abs/1511.05952`
- *Multi-Step Reinforcement Learning: A Unifying Algorithm*, Kristopher de Asis, J Fernando Hernandez-Garcia, G Zacharias Holland, Richard S Sutton: `https://arxiv.org/pdf/1703.01327.pdf`
- *Noisy Networks for Exploration,* by Meire Fortunato, Mohammad Gheshlaghi Azar, Bilal Piot, Jacob Menick, Ian Osband, Alex Graves, Vlad Mnih, Remi Munos, Demis Hassabis, Olivier Pietquin, Charles Blundell, and Shane Legg, arXiv:1706.10295: `https://arxiv.org/abs/1706.10295`

5
Deep Deterministic Policy Gradient

In earlier chapters, you saw the use of **reinforcement learning** (RL) to solve discrete action problems, such as those that arise in Atari games. We will now build on this to tackle continuous, real-valued action problems. Continuous control problems are copious—for example, the motor torque of a robotic arm; the steering, acceleration, and braking of an autonomous car; the wheeled robotic motion on terrain; and the roll, pitch, and yaw controls of a drone. For these problems, we train neural networks in an RL setting to output real-valued actions.

Many continuous control algorithms involve two neural networks—one referred to as the **actor** (policy-based), and the other as the **critic** (value-based)—and therefore, this family of algorithms is referred to as **Actor-Critic algorithms**. The role of the actor is to learn a good policy that can predict good actions for a given state. The role of the critic is to ascertain whether the actor undertook a good action, and to provide feedback that serves as the learning signal for the actor. This is akin to a student-teacher or employee-boss relationship, wherein the student or employee undertakes a task or work, and the role of the teacher or boss is to provide feedback on the quality of the action performed.

The foundation of continuous control RL is through what is called the **policy gradient**, which is an estimate of how much a neural network's weights should be altered so as to maximize the long-term cumulative discounted rewards. Specifically, it uses the **chain rule** and it is an estimate of the gradient that needs to be back-propagated into the actor network for the policy to improve. It is evaluated as an average over a mini-batch of samples. We will cover these topics in this chapter. In particular, we will cover an algorithm called the **Deep Deterministic Policy Gradient** (DDPG), which is a state-of-the-art RL algorithm for continuous control.

Continuous control has many real-world applications. For instance, continuous control can involve the evaluation of the steering, acceleration, and braking of an autonomous car. It can also be applied to determine the torques required for the actuator motors of a robot. Or, it can be used in biomedical applications, where the control could be to determine the muscle movements for a humanoid locomotion. Thus, continuous control problem applications abound.

The following topics will be covered in this chapter:

- Actor-Critic algorithms and policy gradients
- DDPG
- Training and testing DDPG on Pendulum-v0

Technical requirements

In order to successfully complete this chapter, the following items are required:

- Python (2 and above)
- NumPy
- Matplotlib
- TensorFlow (version 1.4 or higher)
- A computer with at least 8 GB of RAM (higher than this is even better!)

Actor-Critic algorithms and policy gradients

In this section, we will cover what Actor-Critic algorithms are. You will also see what policy gradients are and how they are useful to Actor-Critic algorithms.

How do students learn at school? Students normally make a lot of mistakes as they learn. When they do well at learning a task, their teacher provides positive feedback. On the other hand, if students do poorly at a task, the teacher provides negative feedback. This feedback serves as the learning signal for the student to get better at their tasks. This is the crux of Actor-Critic algorithms.

The following is a summary of the steps involved:

- We will have two neural networks—one referred to as the actor, and the other as the critic
- The actor is like the student, as we described previously, and takes an action at a given state
- The critic is like the teacher, as we described previously, and provides feedback for the actor to learn
- Unlike a teacher in a school, the critic network should also be trained from scratch, which makes the problem challenging
- The policy gradient is used to train the actor
- The L2 norm on the Bellman update is used to train the critic

Policy gradient

The **policy gradient** is defined as follows:

$$\nabla_\theta J = \frac{1}{N} \sum_i \nabla_a Q(s, a) \nabla_\theta \pi(s)$$

J is the long-term reward function that needs to be maximized, *θ* is the policy neural network parameters, *N* is the mini-batch size, *Q(s,a)* is the state-action value function, and *π* is the policy. In other words, we compute the gradient of the state-action value function with respect to actions and the gradient of the policy with respect to the network parameters, multiply them, and take an average of them over *N* samples of data from a mini-batch. We can then use this policy gradient in a gradient ascent setting to update the policy parameters. Note that it is essentially a chain rule of calculus that is being used to evaluate the policy gradient.

Deep Deterministic Policy Gradient

We will now delve into the DDPG algorithm, which is a state-of-the-art RL algorithm for continuous control. It was originally published by Google DeepMind in 2016 and has gained a lot of interest in the community, with several new variants proposed thereafter. As was the case in DQN, DDPG also uses target networks for stability. It also uses a replay buffer to reuse past data, and therefore, it is an off-policy RL algorithm.

The `ddpg.py` file is the main file from which we start the training and testing. It will call the training or testing functions, which are present in `TrainOrTest.py`. The `AandC.py` file has the TensorFlow code for the actor and the critic networks. Finally, `replay_buffer.py` stores the samples in a replay buffer by using a deque data structure. We will train the DDPG to learn to hold an inverted pendulum vertically, using OpenAI Gym's Pendulum-v0, which has three states and one continuous action, which is the torque to be applied to hold the pendulum as vertically inverted.

Coding ddpg.py

We will first code the `ddpg.py` file. The steps that are involved are as follows.

We will now summarize the DDPG code:

1. **Importing the required packages**: We will import the required packages and other Python files:

```
import tensorflow as tf
import numpy as np
import gym
from gym import wrappers

import argparse
import pprint as pp
import sys

from replay_buffer import ReplayBuffer
from AandC import *
from TrainOrTest import *
```

2. **Defining the train() function**: We will define the `train()` function. This takes the argument parser object, `args`. We create a TensorFlow session as `sess`. The name of the environment is used to make a Gym environment stored in the `env` object. We also set the random number of seeds and the maximum number of steps for an episode of the environment. We also set the state and action dimensions in `state_dim` and `action_dim`, which take the values of 3 and 1, respectively, for the Pendulum-v0 problem. We then create actor and critic objects, which are instances of the `ActorNetwork` class and the `CriticNetwork` class, respectively, which will be described later, in the `AandC.py file`. We then call the `trainDDPG()` function, which will start the training of the RL agent.

Finally, we save the TensorFlow model by using `tf.train.Saver()` and `saver.save()`:

```
def train(args):

    with tf.Session() as sess:

        env = gym.make(args['env'])
        np.random.seed(int(args['random_seed']))
        tf.set_random_seed(int(args['random_seed']))
        env.seed(int(args['random_seed']))
        env._max_episode_steps = int(args['max_episode_len'])

        state_dim = env.observation_space.shape[0]
        action_dim = env.action_space.shape[0]
        action_bound = env.action_space.high

        actor = ActorNetwork(sess, state_dim, action_dim,
action_bound,
                    float(args['actor_lr']), float(args['tau']),
int(args['minibatch_size']))

        critic = CriticNetwork(sess, state_dim, action_dim,
                    float(args['critic_lr']), float(args['tau']),
float(args['gamma']), actor.get_num_trainable_vars())

        trainDDPG(sess, env, args, actor, critic)

        saver = tf.train.Saver()
        saver.save(sess, "ckpt/model")
        print("saved model ")
```

3. **Defining the test() function**: The `test()` function is defined next. This will be used once we have finished the training and want to test how well our agent is performing. The code is as follows for the `test()` function and is very similar to `train()`. We will restore the saved model from `train()` by using `tf.train.Saver()` and `saver.restore()`. We call the `testDDPG()` function to test the model:

```
def test(args):

    with tf.Session() as sess:

        env = gym.make(args['env'])
        np.random.seed(int(args['random_seed']))
        tf.set_random_seed(int(args['random_seed']))
        env.seed(int(args['random_seed']))
```

```
        env._max_episode_steps = int(args['max_episode_len'])

        state_dim = env.observation_space.shape[0]
        action_dim = env.action_space.shape[0]
        action_bound = env.action_space.high

        actor = ActorNetwork(sess, state_dim, action_dim,
action_bound,
                float(args['actor_lr']), float(args['tau']),
int(args['minibatch_size']))

        critic = CriticNetwork(sess, state_dim, action_dim,
                float(args['critic_lr']), float(args['tau']),
float(args['gamma']), actor.get_num_trainable_vars())

        saver = tf.train.Saver()
        saver.restore(sess, "ckpt/model")

        testDDPG(sess, env, args, actor, critic)
```

4. **Defining the main function**: Finally, the `main` function is as follows. We define an argument parser by using Python's `argparse`. The learning rates for the actor and critic are specified, including the discount factor, `gamma`, and the target network exponential average parameter, `tau`. The buffer size, mini-batch size, and number of episodes are also specified in the argument parser. The environment that we are interested in is Pendulum-v0, and this is also specified in the argument parser.

5. **Calling the train() or test() function, as appropriate**: The mode for running this code is train or test, and it will call the appropriate eponymous function, which we defined previously:

```
if __name__ == '__main__':
    parser = argparse.ArgumentParser(description='provide arguments
for DDPG agent')

    # agent parameters
    parser.add_argument('--actor-lr', help='actor network learning
rate', default=0.0001)
    parser.add_argument('--critic-lr', help='critic network
learning rate', default=0.001)
    parser.add_argument('--gamma', help='discount factor for
Bellman updates', default=0.99)
    parser.add_argument('--tau', help='target update parameter',
default=0.001)
    parser.add_argument('--buffer-size', help='max size of the
replay buffer', default=1000000)
```

```
    parser.add_argument('--minibatch-size', help='size of
minibatch', default=64)

    # run parameters
    parser.add_argument('--env', help='gym env', default='Pendulum-
v0')
    parser.add_argument('--random-seed', help='random seed',
default=258)
    parser.add_argument('--max-episodes', help='max num of
episodes', default=250)
    parser.add_argument('--max-episode-len', help='max length of
each episode', default=1000)
    parser.add_argument('--render-env', help='render gym env',
action='store_true')
    parser.add_argument('--mode', help='train/test',
default='train')
    args = vars(parser.parse_args())
    pp.pprint(args)

    if (args['mode'] == 'train'):
      train(args)
    elif (args['mode'] == 'test'):
      test(args)
```

That's it for ddpg.py.

Coding AandC.py

We will specify the ActorNetwork class and the CriticNetwork class in AandC.py. The steps involved are as follows:

1. **Importing packages**: First, we import the packages:

```
import tensorflow as tf
import numpy as np
import gym
from gym import wrappers
import argparse
import pprint as pp
import sys

from replay_buffer import ReplayBuffer
```

2. **Define initializers for the weights and biases**: Next, we define the weights and biases initializers:

```
winit = tf.contrib.layers.xavier_initializer()
binit = tf.constant_initializer(0.01)
rand_unif =
tf.keras.initializers.RandomUniform(minval=-3e-3,maxval=3e-3)
regularizer = tf.contrib.layers.l2_regularizer(scale=0.0)
```

3. **Defining the ActorNetwork class**: The `ActorNetwork` class is specified as follows. First, it receives parameters as arguments in the `__init__` constructor. We then call `create_actor_network()`, which will return `inputs`, `out`, and `scaled_out` objects. The actor model parameters are stored in `self.network_params` by calling TensorFlow's `tf.trainable_variables()`. We replicate the same for the actor's target network, as well. Note that the target network is required for stability reasons; it is identical to the actor in neural network architecture, albeit the parameters gradually change. The target network parameters are collected and stored in `self.target_network_params` by calling `tf.trainable_variables()` again:

```
class ActorNetwork(object):

    def __init__(self, sess, state_dim, action_dim, action_bound,
learning_rate, tau, batch_size):
        self.sess = sess
        self.s_dim = state_dim
        self.a_dim = action_dim
        self.action_bound = action_bound
        self.learning_rate = learning_rate
        self.tau = tau
        self.batch_size = batch_size

        # actor
        self.state, self.out, self.scaled_out =
self.create_actor_network(scope='actor')

        # actor params
        self.network_params = tf.trainable_variables()

        # target network
        self.target_state, self.target_out, self.target_scaled_out
= self.create_actor_network(scope='act_target')
        self.target_network_params =
tf.trainable_variables()[len(self.network_params):]
```

4. **Defining self.update_target_network_params**: Next, we define
   `self.update_target_network_params`, which will weigh the current actor
   network parameters with `tau` and the target network's parameters with `1-tau`,
   and add them to store them as a TensorFlow operation. We are thus gradually
   updating the target network's model parameters. Note the use of
   `tf.multiply()` for multiplying the weights with `tau` (or `1-tau`, as the case
   may be). We then create a TensorFlow placeholder called `action_gradient` to
   store the gradient of *Q*, with respect to the action, which is to be supplied by the
   critic. We also use `tf.gradients()` to compute the gradient of the output of the
   policy network with respect to the network parameters. Note that we then divide
   by the `batch_size`, in order to average the summation over a mini-batch. This
   gives us the averaged policy gradient, which we can then use to update the actor
   network parameters:

```
# update target using tau and 1-tau as weights
self.update_target_network_params = \
[self.target_network_params[i].assign(tf.multiply(self.network_para
ms[i], self.tau) + tf.multiply(self.target_network_params[i], 1. -
self.tau))
        for i in range(len(self.target_network_params))]

# gradient (this is provided by the critic)
self.action_gradient = tf.placeholder(tf.float32, [None,
self.a_dim])

# actor gradients
self.unnormalized_actor_gradients = tf.gradients(
    self.scaled_out, self.network_params, -self.action_gradient)
self.actor_gradients = list(map(lambda x: tf.div(x,
self.batch_size), self.unnormalized_actor_gradients))
```

5. **Using Adam optimization**: We use Adam optimization to apply the policy
   gradients, in order to optimize the actor's policy:

```
 # adam optimization
        self.optimize =
tf.train.AdamOptimizer(self.learning_rate).apply_gradients(zip(self
.actor_gradients, self.network_params))

        # num trainable vars
        self.num_trainable_vars = len(self.network_params) +
len(self.target_network_params)
```

6. **Defining the create_actor_network() function**: We now define the
   `create_actor_network()` function. We will use a neural network with two
   layers, with `400` and `300` neurons, respectively. The weights are initialized by
   using **Xavier initialization**, and the biases are zeros to begin with. We use
   the `relu` activation function, and also batch normalization, for stability. The final
   output layer has weights initialized with a uniform distribution and a `tanh`
   activation function in order to keep it bounded. For the Pendulum-v0 problem,
   the actions are bounded in the range [*-2,2*], and since `tanh` is bounded in the
   range [*-1,1*], we need to multiply the output by two to scale accordingly; this is
   done by using `tf.multiply()`, where `action_bound = 2` for the inverted
   pendulum problem:

```
def create_actor_network(self, scope):
    with tf.variable_scope(scope, reuse=tf.AUTO_REUSE):
        state = tf.placeholder(name='a_states', dtype=tf.float32,
shape=[None, self.s_dim])
        net = tf.layers.dense(inputs=state, units=400,
activation=None, kernel_initializer=winit, bias_initializer=binit,
name='anet1')
        net = tf.nn.relu(net)

        net = tf.layers.dense(inputs=net, units=300,
activation=None, kernel_initializer=winit, bias_initializer=binit,
name='anet2')
        net = tf.nn.relu(net)

        out = tf.layers.dense(inputs=net, units=self.a_dim,
activation=None, kernel_initializer=rand_unif,
bias_initializer=binit, name='anet_out')
        out = tf.nn.tanh(out)
        scaled_out = tf.multiply(out, self.action_bound)
        return state, out, scaled_out
```

7. **Define the actor functions**: Finally, we have the remaining functions that
   are required to complete the `ActorNetwork` class. We will define `train()`,
   which will run a session on `self.optimize`; the `predict()` function runs a
   session on `self.scaled_out`, that is, the output of the `ActorNetwork`; the
   `predict_target()` function will run a session on `self.target_scaled_out`,
   that is, the output action of the actor's target network. Then,
   `update_target_network()` will run a session on
   `self.update_target_network_params`, which will perform the weighted
   average of the network parameters.

Finally, the `get_num_trainable_vars()` function returns a count of the number of trainable variables:

```
def train(self, state, a_gradient):
        self.sess.run(self.optimize, feed_dict={self.state: state,
self.action_gradient: a_gradient})

def predict(self, state):
        return self.sess.run(self.scaled_out, feed_dict={
            self.state: state})

def predict_target(self, state):
        return self.sess.run(self.target_scaled_out, feed_dict={
            self.target_state: state})

def update_target_network(self):
        self.sess.run(self.update_target_network_params)

def get_num_trainable_vars(self):
        return self.num_trainable_vars
```

8. **Defining CriticNetwork class**: We will now define the `CriticNetwork` class. Similar to `ActorNetwork`, we receive the model hyperparameters as arguments. We then call the `create_critic_network()` function, which will return `inputs`, `action`, and `out`. We also create the target network for the critic by calling `create_critic_network()` again:

```
class CriticNetwork(object):

    def __init__(self, sess, state_dim, action_dim, learning_rate,
tau, gamma, num_actor_vars):
        self.sess = sess
        self.s_dim = state_dim
        self.a_dim = action_dim
        self.learning_rate = learning_rate
        self.tau = tau
        self.gamma = gamma

        # critic
        self.state, self.action, self.out =
self.create_critic_network(scope='critic')

        # critic params
        self.network_params =
tf.trainable_variables()[num_actor_vars:]

        # target Network
```

```
        self.target_state, self.target_action, self.target_out =
    self.create_critic_network(scope='crit_target')

        # target network params
        self.target_network_params =
    tf.trainable_variables()[(len(self.network_params) +
    num_actor_vars):]
```

9. **Critic target network**: Similar to the actor's target, the critic's target network is also updated by using weighted averaging. We then create a TensorFlow placeholder called `predicted_q_value`, which is the target value. We then define the L2 norm in `self.loss`, which is the quadratic error on the Bellman residual. Note that `self.out` is the $Q(s,a)$ that we saw earlier, and `predicted_q_value` is the $r + \gamma Q(s',a')$ in the Bellman equation. Again, we use the Adam optimizer to minimize this L2 loss function. We then evaluate the gradient of $Q(s,a)$ with respect to the actions by calling `tf.gradients()`, and we store this in `self.action_grads`. This gradient is used later in the computation of the policy gradients:

```
# update target using tau and 1 - tau as weights
        self.update_target_network_params = \
[self.target_network_params[i].assign(tf.multiply(self.network_para
ms[i], self.tau) \
        + tf.multiply(self.target_network_params[i], 1. -
self.tau))
                for i in range(len(self.target_network_params))]

        # network target (y_i in the paper)
        self.predicted_q_value = tf.placeholder(tf.float32, [None,
1])

        # adam optimization; minimize L2 loss function
        self.loss = tf.reduce_mean(tf.square(self.predicted_q_value
- self.out))
        self.optimize =
tf.train.AdamOptimizer(self.learning_rate).minimize(self.loss)

        # gradient of Q w.r.t. action
        self.action_grads = tf.gradients(self.out, self.action)
```

10. **Defining create_critic_network()**: Next, we will define the create_critic_network() function. The critic network is also similar to the actor in architecture, except that it takes both the states and the actions as input. There are two hidden layers, with 400 and 300 neurons, respectively. The last output layer has only one neuron, that is, is the *Q(s,a)* state-action value function. Note that the last layer has no activation function, as Q(s,a) is, in theory, unbounded:

```
def create_critic_network(self, scope):
        with tf.variable_scope(scope, reuse=tf.AUTO_REUSE):
            state = tf.placeholder(name='c_states',
dtype=tf.float32, shape=[None, self.s_dim])
            action = tf.placeholder(name='c_action',
dtype=tf.float32, shape=[None, self.a_dim])

            net = tf.concat([state, action],1)

            net = tf.layers.dense(inputs=net, units=400,
activation=None, kernel_initializer=winit, bias_initializer=binit,
name='cnet1')
            net = tf.nn.relu(net)

            net = tf.layers.dense(inputs=net, units=300,
activation=None, kernel_initializer=winit, bias_initializer=binit,
name='cnet2')
            net = tf.nn.relu(net)

            out = tf.layers.dense(inputs=net, units=1,
activation=None, kernel_initializer=rand_unif,
bias_initializer=binit, name='cnet_out')
            return state, action, out
```

11. Finally, the functions required to complete the CriticNetwork are as follows. These are similar to the ActorNetwork, so we do not elaborate further for brevity. One difference, however, is the action_gradients() function, which is the gradient of *Q(s,a)* with respect to the actions, which is computed by the critic and supplied to the actor, to be used in the evaluation of the policy gradients:

```
def train(self, state, action, predicted_q_value):
        return self.sess.run([self.out, self.optimize],
feed_dict={self.state: state, self.action: action,
self.predicted_q_value: predicted_q_value})

def predict(self, state, action):
        return self.sess.run(self.out, feed_dict={self.state:
```

```
state, self.action: action})

def predict_target(self, state, action):
        return self.sess.run(self.target_out,
feed_dict={self.target_state: state, self.target_action: action})

def action_gradients(self, state, actions):
        return self.sess.run(self.action_grads,
feed_dict={self.state: state, self.action: actions})

    def update_target_network(self):
        self.sess.run(self.update_target_network_params)
```

That's it for `AandC.py`.

Coding TrainOrTest.py

The `trainDDPG()` and `testDDPG()` functions that we used earlier will now be defined in `TrainOrTest.py`. The steps that are involved are as follows:

1. **Import packages and functions**: The `TrainOrTest.py` file starts with the importing of the packages and other Python files:

   ```
   import tensorflow as tf
   import numpy as np
   import gym
   from gym import wrappers

   import argparse
   import pprint as pp
   import sys

   from replay_buffer import ReplayBuffer
   from AandC import *
   ```

2. **Define the trainDDPG() function**: Next, we define the `trainDDPG()` function. First, we initialize all of the networks by calling a `sess.run()` on `tf.global_variables_initializer()`. Then, we initialize the target network weights and the replay buffer. Then, we start the main loop over the training episodes. Inside of this loop, we reset the environment (Pendulum-v0, in our case) and also start the loop over time steps for each episode (recall that each episode has a `max_episode_len` number of time steps).

The actor's policy is sampled to obtain the action for the current state. We feed this action into `env.step()`, which takes one time step of this action and, in the process, moves to the next state, `s2`. The environment also gives this a reward, `r`, and information on whether the episode is terminated is stored in the Boolean variable `terminal`. We add the tuple (`state`, `action`, `reward`, `terminal`, `new state`) to the replay buffer for sampling later and for training:

```
def trainDDPG(sess, env, args, actor, critic):

    sess.run(tf.global_variables_initializer())


    # Initialize target networks
    actor.update_target_network()
    critic.update_target_network()

    # Initialize replay memory
    replay_buffer = ReplayBuffer(int(args['buffer_size']),
int(args['random_seed']))

    # start training on episodes
    for i in range(int(args['max_episodes'])):

        s = env.reset()

        ep_reward = 0
        ep_ave_max_q = 0

        for j in range(int(args['max_episode_len'])):

            if args['render_env']:
                env.render()

            a = actor.predict(np.reshape(s, (1, actor.s_dim)))
            s2, r, terminal, info = env.step(a[0])

            replay_buffer.add(np.reshape(s, (actor.s_dim,)),
np.reshape(a, (actor.a_dim,)), r,
                                terminal, np.reshape(s2,
(actor.s_dim,)))
```

3. **Sample a mini-batch of data from the replay buffer**: Once we have more than the mini-batch size of samples in the replay buffer, we sample a mini-batch of data from the buffer. For the subsequent state, s2, we use the critic's target network to compute the target Q value and store it in target_q. Note the use of the critic's target and not the critic—this is done for stability reasons. We then use the Bellman equation to evaluate the target, y_i, which is computed as $r + \gamma Q$ for non-Terminal time steps and as r for Terminal steps:

```
# sample from replay buffer
        if replay_buffer.size() > int(args['minibatch_size']):
            s_batch, a_batch, r_batch, t_batch, s2_batch =
            replay_buffer.sample_batch(int(args['minibatch
            _size']))

            # Calculate target q
            target_q = critic.predict_target(s2_batch,
                    actor.predict_target(s2_batch))

            y_i = []
            for k in range(int(args['minibatch_size'])):
                if t_batch[k]:
                    y_i.append(r_batch[k])
                else:
                    y_i.append(r_batch[k] + critic.gamma *
                                    target_q[k])
```

4. **Use the preceding to train the actor and critic**: We then train the critic for one step on the mini-batch by calling critic.train(). Then, we compute the gradient of Q with respect to the action by calling critic.action_gradients() and we store it in grads; note that this action gradient is used to compute the policy gradient, as we mentioned previously. We then train the actor for one step by calling actor.train() and passing grads as an argument, along with the state that we sampled from the replay buffer. Finally, we update the actor and critic target networks by calling the appropriate functions for the actor and critic objects:

```
# Update critic
            predicted_q_value, _ = critic.train(s_batch,
    a_batch, np.reshape(y_i, (int(args['minibatch_size']), 1)))

            ep_ave_max_q += np.amax(predicted_q_value)

            # Update the actor policy using gradient
            a_outs = actor.predict(s_batch)
            grads = critic.action_gradients(s_batch, a_outs)
```

```
actor.train(s_batch, grads[0])

# update target networks
actor.update_target_network()
critic.update_target_network()
```

The new state, s2, is assigned to the current state, s, as we proceed to the next time step. If the episode has terminated, we print the episode reward and other observations on the screen, and we write them into a text file called pendulum.txt for later analysis. We also break out of the inner for loop, as the episode has terminated:

```
s = s2
ep_reward += r

if terminal:
    print('| Episode: {:d} | Reward: {:d} | Qmax: {:.4f}'.format(i,
        int(ep_reward), (ep_ave_max_q / float(j))))
    f = open("pendulum.txt", "a+")
    f.write(str(i) + " " + str(int(ep_reward)) + " " +
        str(ep_ave_max_q / float(j)) + '\n')
    break
```

5. **Defining testDDPG()**: This concludes the trainDDPG() function. We will now present the testDDPG() function that is used to test how well our model is performing. The testDDPG() function is more or less the same as trainDDPG(), except that we do not have a replay buffer and we do not train the neural networks. Like before, we have two for loops—the outer one for episodes, and the inner loop over time steps for each episode. We sample actions from the trained actor's policy by using actor.predict() and use it to evolve the environment by using env.step(). Finally, we terminate the episode if terminal == True:

```
def testDDPG(sess, env, args, actor, critic):


    # test for max_episodes number of episodes
    for i in range(int(args['max_episodes'])):

        s = env.reset()

        ep_reward = 0
        ep_ave_max_q = 0

        for j in range(int(args['max_episode_len'])):
```

```
                      if args['render_env']:
                          env.render()

                      a = actor.predict(np.reshape(s, (1, actor.s_dim)))

                      s2, r, terminal, info = env.step(a[0])

                      s = s2
                      ep_reward += r

                      if terminal:
                          print('| Episode: {:d} | Reward: {:d} |'.format(i,
                                  int(ep_reward)))
                          break
```

This concludes `TrainOrTest.py`.

Coding replay_buffer.py

We will use the deque data structure for storing our replay buffer. The steps that are involved are as follows:

1. **Import the packages**: First, we import the required packages.
2. **Define the ReplayBuffer class**: We then define the `ReplayBuffer` class, with the arguments passed to the `__init__()` constructor. The `self.buffer = deque()` function is the instance of the data structure to store the data in a queue:

```
from collections import deque
import random
import numpy as np

class ReplayBuffer(object):

    def __init__(self, buffer_size, random_seed=258):
        self.buffer_size = buffer_size
        self.count = 0
        self.buffer = deque()
        random.seed(random_seed)
```

3. **Define the add and size functions**: We then define the `add()` function to add the experience as a tuple (`state`, `action`, `reward`, `terminal`, `new state`). The `self.count` function keeps a count of the number of samples we have in the replay buffer. If this count is less than the replay buffer size (`self.buffer_size`), we append the current experience to the buffer and increment the count. On the other hand, if the count is equal to (or greater than) the buffer size, we discard the old samples from the buffer by calling `popleft()`, which is a built-in function of deque. Then, we add the experience to the replay buffer; the count need not be incremented, as we discarded one old data sample in the replay buffer and replaced it with the new data sample or experience, so the total number of samples in the buffer remains the same. We also define the `size()` function to obtain the current size of the replay buffer:

```
def add(self, s, a, r, t, s2):
    experience = (s, a, r, t, s2)
    if self.count < self.buffer_size:
        self.buffer.append(experience)
        self.count += 1
    else:
        self.buffer.popleft()
        self.buffer.append(experience)

def size(self):
    return self.count
```

4. **Define the sample_batch and clear functions**: Next, we define the `sample_batch()` function to sample a `batch_size` number of samples from the replay buffer. If the count of the number of samples in the buffer is less than the `batch_size`, we sample count the number of samples from the buffer. Otherwise, we sample the `batch_size` number of samples from the replay buffer. Then, we convert these samples to `NumPy` arrays and return them. Lastly, the `clear()` function is used to completely clear the reply buffer and make it empty:

```
def sample_batch(self, batch_size):
    batch = []

    if self.count < batch_size:
        batch = random.sample(self.buffer, self.count)
    else:
        batch = random.sample(self.buffer, batch_size)

    s_batch = np.array([_[0] for _ in batch])
    a_batch = np.array([_[1] for _ in batch])
```

```
        r_batch = np.array([_[2] for _ in batch])
        t_batch = np.array([_[3] for _ in batch])
        s2_batch = np.array([_[4] for _ in batch])

        return s_batch, a_batch, r_batch, t_batch, s2_batch

def clear(self):
        self.buffer.clear()
        self.count = 0
```

That concludes the code for the DDPG. We will now test it.

Training and testing the DDPG on Pendulum-v0

We will now train the preceding DDPG code on Pendulum-v0. To train the DDPG agent, simply type the following in the command line at the same level as the rest of the code:

```
python ddpg.py
```

This will start the training:

```
{'actor_lr': 0.0001,
 'buffer_size': 1000000,
 'critic_lr': 0.001,
 'env': 'Pendulum-v0',
 'gamma': 0.99,
 'max_episode_len': 1000,
 'max_episodes': 250,
 'minibatch_size': 64,
 'mode': 'train',
 'random_seed': 258,
 'render_env': False,
 'tau': 0.001}
.
.
.
2019-03-03 17:23:10.529725: I
tensorflow/stream_executor/cuda/cuda_diagnostics.cc:300] kernel version
seems to match DSO: 384.130.0
| Episode: 0 | Reward: -7981 | Qmax: -6.4859
| Episode: 1 | Reward: -7466 | Qmax: -10.1758
| Episode: 2 | Reward: -7497 | Qmax: -14.0578
```

Once the training is complete, you can also test the trained DDPG agent, as follows:

```
python ddpg.py --mode test
```

We can also plot the episodic rewards during training by using the following code:

```
import numpy as np
import matplotlib.pyplot as plt

data = np.loadtxt('pendulum.txt')

plt.plot(data[:,0], data[:,1])
plt.xlabel('episode number', fontsize=12)
plt.ylabel('episode reward', fontsize=12)
#plt.show()
plt.savefig("ddpg_pendulum.png")
```

The plot is presented as follows:

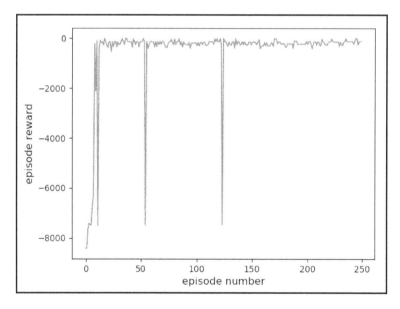

Figure 1: Plot showing the episode rewards during training for the Pendulum-v0 problem, using the DDPG

As you can see, the DDPG agent has learned the problem very well. The maximum rewards are slightly negative, and this is the best for this problem.

Summary

In this chapter, we were introduced to our first continuous actions RL algorithm, DDPG, which also happens to be the first Actor-Critic algorithm in this book. DDPG is an off-policy algorithm, as it uses a replay buffer. We also covered the use of policy gradients to update the actor, and the use of the L2 norm to update the critic. Thus, we have two different neural networks. The actor learns the policy and the critic learns to evaluate the actor's policy, thereby providing a learning signal to the actor. You saw how to compute the gradient of the state-action value, $Q(s,a)$, with respect to the action, and also the gradient of the policy, both of which are combined to evaluate the policy gradient, which is then used to update the actor. We trained the DDPG on the inverted pendulum problem, and the agent learned it very well.

We have come a long way in this chapter. You have learned about Actor-Critic algorithms and how to code your first continuous control RL algorithm. In the next chapter, you will learn about the **A3C algorithm**, which is an on-policy deep RL algorithm.

Questions

1. Is the DDPG an on-policy or off-policy algorithm?
2. We used the same neural network architectures for both the actor and the critic. Is this required, or can we choose different neural network architectures for the actor and the critic?
3. Can we use the DDPG for Atari Breakout?
4. Why are the biases of the neural networks initialized to small positive values?
5. This is left as an exercise: Can you modify the code in this chapter to train an agent to learn InvertedDoublePendulum-v2, which is more challenging than the Pendulum-v0 that you saw in this chapter?
6. Here is another exercise: Vary the neural network architecture and check whether the agent can learn the Pendulum-v0 problem. For instance, keep decreasing the number of neurons in the first hidden layer with the values 400, 100, 25, 10, 5, and 1, and check how the agent performs for the different number of neurons in the first hidden layer. If the number of neurons is too small, it can lead to information bottlenecks, where the input of the network is not sufficiently represented; that is, the information is lost as we go deeper into the neural network. Do you observe this effect?

Further reading

- *Continuous control with deep reinforcement learning,* by *Timothy P. Lillicrap, Jonathan J. Hunt, Alexander Pritzel, Nicolas Heess, Tom Erez, Yuval Tassa, David Silver,* and *Daan Wierstra,* original DDPG paper from *DeepMind,* arXiv:1509.02971: `https://arxiv.org/abs/1509.02971`

Asynchronous Methods - A3C and A2C

6

We looked at the DDPG algorithm in the previous chapter. One of the main drawbacks of the DDPG algorithm (as well as the DQN algorithm that we saw earlier) is the use of a replay buffer to obtain independent and identically distributed samples of data for training. Using a replay buffer consumes a lot of memory, which is not desirable for robust RL applications. To overcome this problem, researchers at Google DeepMind came up with an on-policy algorithm called **Asynchronous Advantage Actor Critic** (**A3C**). A3C does not use a replay buffer; instead, it uses parallel worker processors, where different instances of the environment are created and the experience samples are collected. Once a finite and fixed number of samples are collected, they are used to compute the policy gradients, which are asynchronously sent to a central processor that updates the policy. This updated policy is then sent back to the worker processors. The use of parallel processors to experience different scenarios of the environment gives rise to independent and identically distributed samples that can be used to train the policy. This chapter will cover A3C, and will also briefly touch upon a variant of it called the **Advantage Actor Critic** (**A2C**).

The following topics will be covered in this chapter:

- The A3C algorithm
- The A3C algorithm applied to CartPole
- The A3C algorithm applied to LunarLander
- The A2C algorithm

In this chapter, you will learn about the A3C and A2C algorithms, as well as how to code them using Python and TensorFlow. We will also apply the A3C algorithm to solving two OpenAI Gym problems: CartPole and LunarLander.

Technical requirements

To successfully complete this chapter, some knowledge of the following will be of great help:

- TensorFlow (version 1.4 or higher)
- Python (version 2 or 3)
- NumPy

The A3C algorithm

As we mentioned earlier, we have parallel workers in A3C, and each worker will compute the policy gradients and pass them on to the central (or master) processor. The A3C paper also uses the `advantage` function to reduce variance in the policy gradients. The `loss` functions consist of three losses, which are weighted and added; they include the value loss, the policy loss, and an entropy regularization term. The value loss, L_v, is an L2 loss of the state value and the target value, with the latter computed as a discounted sum of the rewards. The policy loss, L_p, is the product of the logarithm of the policy distribution and the `advantage` function, A. The entropy regularization, L_e, is the Shannon entropy, which is computed as the product of the policy distribution and its logarithm, with a minus sign included. The entropy regularization term is like a bonus for exploration; the higher the entropy, the better regularized the ensuing policy is. These three terms are weighted as 0.5, 1, and -0.005, respectively.

Loss functions

The value loss is computed as the weighted sum of three loss terms: the value loss, L_v, the policy loss, L_p, and the entropy regularization term, L_e, which are evaluated as follows:

$$L_v = \sum \left(V^{target} - V(s_t) \right)^2$$
$$L_p = - \sum log\pi_\theta(a_t|s_t)A(s_t, a_t)$$
$$L_e = - \sum \pi_\theta(a_t|s_t)log\pi_\theta(a_t|s_t)$$
$$L = 0.5L_v + L_p - 0.005L_e$$

L is the total loss, which has to be minimized. Note that we would like to maximize the `advantage` function, so we have a minus sign in L_p, as we are minimizing L. Likewise, we would like to maximize the entropy term, and since we are minimizing L, we have a minus sign in the term $-0.005 L_e$ in L.

CartPole and LunarLander

In this section, we will apply A3C to OpenAI Gym's CartPole and LunarLander.

CartPole

CartPole consists of a vertical pole on a cart that needs to be balanced by moving the cart either to the left or to the right. The state dimension is four and the action dimension is two for CartPole.

Check out the following link for more details on CartPole: `https://gym.openai.com/envs/CartPole-v0/`.

LunarLander

LunarLander, as the name suggests, involves the landing of a lander on the lunar surface. For example, when Apollo 11's Eagle lander touched down on the moon's surface in 1969, the astronauts Neil Armstrong and Buzz Aldrin had to control the rocket thrusters during the final phase of the descent and safely land the spacecraft on the surface. After this, of course, Armstrong walked on the moon and remarked the now famous sentence: "*One small step for a man, one giant leap for mankind*". In LunarLander, there are two yellow flags on the lunar surface, and the goal is to land the spacecraft between these flags. Fuel in the lander is infinite, unlike the case in Apollo 11's Eagle lander. The state dimension is eight and the action dimension is four for LunarLander, with the four actions being do nothing, fire the left thruster, fire the main thruster, or fire the right thruster.

Check out the following link for a schematic of the environment: `https://gym.openai.com/envs/LunarLander-v2/`.

The A3C algorithm applied to CartPole

Here, we will code A3C in TensorFlow and apply it so that we can train an agent to learn the CartPole problem. The following code files will be required to code:

- `cartpole.py`: This will start the training or testing process
- `a3c.py`: This is where the A3C algorithm is coded
- `utils.py`: This includes utility functions

Coding cartpole.py

We will now code `cartpole.py`. Follow these steps to get started:

1. First, we import the packages:

```
import numpy as np
import matplotlib.pyplot as plt
import tensorflow as tf
import gym
import os
import threading
import multiprocessing

from random import choice
from time import sleep
from time import time

from a3c import *
from utils import *
```

2. Next, we set the parameters for the problem. We only need to train for 200 episodes (yes, CartPole is an easy problem!). We set the discount factor gamma to 0.99. The state and action dimensions are 4 and 2, respectively, for CartPole. If you want to load a pre-trained model and resume training, set `load_model` to `True`; for fresh training from scratch, set this to `False`. We will also set the `model_path`:

```
max_episode_steps = 200
gamma = 0.99
s_size = 4
a_size = 2
load_model = False
model_path = './model'
```

3. We reset the TensorFlow graph and also create a directory for storing our model. We will refer to the master processor as CPU 0. Worker threads have non-zero CPU numbers. The master processor will undertake the following: first, it will create a count of global variables in the global_episodes object. The total number of worker threads will be stored in num_workers, and we can use Python's multiprocessing library to obtain the number of available processors in our system by calling cpu_count(). We will use the Adam optimizer and store it in an object called trainer, along with an appropriate learning rate. We will later define an actor critic class called AC, so we must first create a master network object of the type AC class, called master_network. We will also pass the appropriate arguments to the class' constructor. Then, for each worker thread, we will create a separate instance of the CartPole environment and an instance of a Worker class, which will soon be defined. Finally, for saving the model, we will also create a TensorFlow saver:

```
tf.reset_default_graph()

if not os.path.exists(model_path):
    os.makedirs(model_path)


with tf.device("/cpu:0"):

    # keep count of global episodes
    global_episodes =
tf.Variable(0,dtype=tf.int32,name='global_episodes',trainable=False
)

    # number of worker threads
    num_workers = multiprocessing.cpu_count()

    # Adam optimizer
    trainer = tf.train.AdamOptimizer(learning_rate=2e-4,
use_locking=True)
    # global network
    master_network = AC(s_size,a_size,'global',None)
    workers = []
    for i in range(num_workers):
        env = gym.make('CartPole-v0')
workers.append(Worker(env,i,s_size,a_size,trainer,model_path,global
_episodes))

    # tf saver
    saver = tf.train.Saver(max_to_keep=5)
```

4. We then start the TensorFlow session. Inside it, we create a TensorFlow coordinator for the different workers. Then, we either load or restore a pre-trained model or run `tf.global_variables_initializer()` to assign initial values for all the weights and biases:

```
with tf.Session() as sess:

    # tf coordinator for threads
    coord = tf.train.Coordinator()

    if load_model == True:
        print ('Loading Model...')
        ckpt = tf.train.get_checkpoint_state(model_path)
        saver.restore(sess,ckpt.model_checkpoint_path)
    else:
        sess.run(tf.global_variables_initializer())
```

5. Then, we start the `worker_threads`. Specifically, we call the `work()` function, which is part of the `Worker()` class (to be defined soon). `threading.Thread()` will assign one thread to each `worker`. By calling `start()`, we initiate the `worker` thread. In the end, we need to join the threads so that they wait until all the threads finish before they terminate:

```
# start the worker threads
worker_threads = []
for worker in workers:
    worker_work = lambda: worker.work(max_episode_steps, gamma,
sess, coord,saver)
    t = threading.Thread(target=(worker_work))
    t.start()
    worker_threads.append(t)
coord.join(worker_threads)
```

 You can find out more about TensorFlow coordinators at `https://www.tensorflow.org/api_docs/python/tf/train/Coordinator`.

Coding a3c.py

We will now code `a3c.py`. This involves the following steps:

1. Import the packages
2. Set the initializers for weights and biases

3. Define the `AC` class
4. Define the `Worker` class

First, we need to import the necessary packages:

```
import numpy as np
import matplotlib.pyplot as plt
import tensorflow as tf
import gym
import threading
import multiprocessing

from random import choice
from time import sleep
from time import time
from threading import Lock

from utils import *
```

Then, we need to set the initializers for the weights and biases; specifically, we use the Xavier initializer for the weights and zero bias. For the last output layer of the network, the weights are uniform random numbers within a specified range:

```
xavier = tf.contrib.layers.xavier_initializer()
bias_const = tf.constant_initializer(0.05)
rand_unif = tf.keras.initializers.RandomUniform(minval=-3e-3,maxval=3e-3)
regularizer = tf.contrib.layers.l2_regularizer(scale=5e-4)
```

The AC class

We will now describe the `AC` class, which is also part of `a3c.py`. We define the constructor of the `AC` class with an input placeholder, two fully connected hidden layers with 256 and 128 neurons, respectively, and the `elu` activation function. This is followed by the policy network with the `softmax` activation, since our actions our discrete for CartPole. In addition, we also have a value network with no activation function. Note that we share the same hidden layers for both the policy and value, unlike in past examples:

```
class AC():
    def __init__(self,s_size,a_size,scope,trainer):
        with tf.variable_scope(scope):
            self.inputs =
tf.placeholder(shape=[None,s_size],dtype=tf.float32)
            # 2 FC layers
            net = tf.layers.dense(self.inputs, 256, activation=tf.nn.elu,
kernel_initializer=xavier, bias_initializer=bias_const,
```

```
kernel_regularizer=regularizer)
            net = tf.layers.dense(net, 128, activation=tf.nn.elu,
kernel_initializer=xavier, bias_initializer=bias_const,
kernel_regularizer=regularizer)
            # policy
            self.policy = tf.layers.dense(net, a_size,
activation=tf.nn.softmax, kernel_initializer=xavier,
bias_initializer=bias_const)

            # value
            self.value = tf.layers.dense(net, 1, activation=None,
kernel_initializer=rand_unif, bias_initializer=bias_const)
```

For `worker` threads, we need to define the `loss` functions. Thus, when the TensorFlow scope is not `global`, we define an actions placeholder, as well as its one-hot representation; we also define placeholders for the `target` value and `advantage` functions. We then compute the product of the policy distribution and the one-hot actions, sum them, and store them in the `policy_times_a` object. Then, we combine these terms to construct the `loss` functions, as we mentioned previously. We compute the sum over the batch of the L2 loss for value; the Shannon entropy as the policy distribution multiplied with its logarithm, with a minus sign; the policy loss as the product of the logarithm of the policy distribution; and the `advantage` function, summed over the batch of samples. Finally, we use the appropriate weights to combine these losses to compute the total loss, which is stored in `self.loss`:

```
# only workers need tf operations for loss functions and gradient updating
        if scope != 'global':
            self.actions = tf.placeholder(shape=[None],dtype=tf.int32)
            self.actions_onehot =
tf.one_hot(self.actions,a_size,dtype=tf.float32)
            self.target_v =
tf.placeholder(shape=[None],dtype=tf.float32)
            self.advantages =
tf.placeholder(shape=[None],dtype=tf.float32)

            self.policy_times_a = tf.reduce_sum(self.policy *
self.actions_onehot, [1])

            # loss
            self.value_loss = 0.5 *
tf.reduce_sum(tf.square(self.target_v - tf.reshape(self.value,[-1])))
            self.entropy = - tf.reduce_sum(self.policy *
tf.log(self.policy + 1.0e-8))
            self.policy_loss = -
tf.reduce_sum(tf.log(self.policy_times_a + 1.0e-8) * self.advantages)
```

```
        self.loss = 0.5 * self.value_loss + self.policy_loss -
self.entropy * 0.005
```

As you saw in the previous chapter, we use `tf.gradients()` to compute the policy gradients; specifically, we compute the gradients of the `loss` function with respect to the local network variables, with the latter obtained from `tf.get_collection()`. To mitigate the problem of exploding gradients, we clip the gradients to a magnitude of `40.0` using TensorFlow's `tf.clip_by_global_norm()` function. We can then collect the network parameters of the global network using `tf.get_collection()` with a scope of `global` and apply the gradients in the Adam optimizer by using `apply_gradients()`. This will compute the policy gradients:

```
# get gradients from local networks using local losses; clip them to avoid
exploding gradients
local_vars = tf.get_collection(tf.GraphKeys.TRAINABLE_VARIABLES, scope)
self.gradients = tf.gradients(self.loss,local_vars)
self.var_norms = tf.global_norm(local_vars)
grads,self.grad_norms = tf.clip_by_global_norm(self.gradients,40.0)
# apply local gradients to global network using tf.apply_gradients()
global_vars = tf.get_collection(tf.GraphKeys.TRAINABLE_VARIABLES, 'global')
self.apply_grads = trainer.apply_gradients(zip(grads,global_vars))
```

The Worker() class

We will now describe the `Worker()` class, which each worker thread will use. First, we define the __init__() constructor for the class. Inside of it, we define the worker name, the number, the model path, the Adam optimizer, the count of global episodes, and the operator to increment it:

```
class Worker():
    def
__init__(self,env,name,s_size,a_size,trainer,model_path,global_episodes):
        self.name = "worker_" + str(name)
        self.number = name
        self.model_path = model_path
        self.trainer = trainer
        self.global_episodes = global_episodes
        self.increment = self.global_episodes.assign_add(1)
```

We also create the local instance of the AC class, with the appropriate arguments passed in. We then create a TensorFlow operation to copy the model parameters from global to local. We also create a NumPy identity matrix with ones on the diagonal, as well as an environment object:

```
# local copy of the AC network
self.local_AC = AC(s_size,a_size,self.name,trainer)

# tensorflow op to copy global params to local network
self.update_local_ops = update_target_graph('global',self.name)
self.actions = np.identity(a_size,dtype=bool).tolist()
self.env = env
```

Next, we create the train() function, which is the most important part of the Worker class. The states, actions, rewards, next states, or observations and values are obtained from the experience list that's received as an argument by the function. We computed the discounted sum over the rewards by using a utility function called discount(), which we will define soon. Similarly, the advantage function is also discounted:

```
# train function
    def train(self,experience,sess,gamma,bootstrap_value):
        experience = np.array(experience)
        observations = experience[:,0]
        actions = experience[:,1]
        rewards = experience[:,2]
        next_observations = experience[:,3]
        values = experience[:,5]
        # discounted rewards
        self.rewards_plus = np.asarray(rewards.tolist() +
[bootstrap_value])
        discounted_rewards = discount(self.rewards_plus,gamma)[:-1]

        # value
        self.value_plus = np.asarray(values.tolist() + [bootstrap_value])

        # advantage function
        advantages = rewards + gamma * self.value_plus[1:] -
self.value_plus[:-1]
        advantages = discount(advantages,gamma)
```

We then update the global network parameters by calling the TensorFlow operations that we defined earlier, along with the required input to the placeholders that are passed using TensorFlow's `feed_dict` function. Note that since we have multiple worker threads performing this update on the master parameters, we need to avoid conflict. In other words, only one thread can update the master network parameters at a given time instant; two or more threads doing this update at the same time will not update the global parameters one after the other, and it can cause problems if one thread updates the global parameters while the other thread is in the process of updating the same. This means that the former's update will be overwritten by the latter, which we do not desire. This is accomplished using Python's threading library's `Lock()` function. We create an instance of `Lock()` called `lock`. `lock.acquire()` will grant access to the current thread only, which will perform the update, after which it will release the lock using `lock.release()`. Finally, we return the losses from the function:

```
# lock for updating global params
lock = Lock()
lock.acquire()

# update global network params
fd = {self.local_AC.target_v:discounted_rewards,
self.local_AC.inputs:np.vstack(observations),
self.local_AC.actions:actions, self.local_AC.advantages:advantages}
value_loss, policy_loss, entropy, _, _, _ =
sess.run([self.local_AC.value_loss, self.local_AC.policy_loss,
self.local_AC.entropy, self.local_AC.grad_norms, self.local_AC.var_norms,
self.local_AC.apply_grads], feed_dict=fd)

# release lock
lock.release()

return value_loss / len(experience), policy_loss / len(experience), entropy
/ len(experience)
```

Now, we need to define the workers' `work()` function. We first obtain the global episode count and set `total_steps` to zero. Then, inside a TensorFlow session, while the threads are still coordinated, we copy the global parameters to the local network using `self.update_local_ops`. We then start an episode. Since the episode hasn't been terminated, we obtain the policy distribution and store it in `a_dist`. We sample an action from this distribution using NumPy's `random.choice()` function. This action, `a`, is fed into the environment's `step()` function to obtain the new state, the reward, and the Terminal Boolean. We can shape the reward by dividing it by `100.0`.

The experience is stored in the local buffer, called `episode_buffer`. We also add the reward to `episode_reward`, and we increment the `total_steps` count, as well as `episode_step_count`:

```python
# worker's work function
def work(self,max_episode_steps, gamma, sess, coord, saver):
    episode_count = sess.run(self.global_episodes)
    total_steps = 0
    print ("Starting worker " + str(self.number))

        with sess.as_default(), sess.graph.as_default():
            while not coord.should_stop():
                # copy global params to local network
                sess.run(self.update_local_ops)

                # lists for book keeping
                episode_buffer = []
                episode_values = []
                episode_frames = []

                episode_reward = 0
                episode_step_count = 0
                d = False
                s = self.env.reset()
                episode_frames.append(s)

                while not d:
                    # action and value
                    a_dist, v =
sess.run([self.local_AC.policy,self.local_AC.value],
feed_dict={self.local_AC.inputs:[s]})
                    a = np.random.choice(np.arange(len(a_dist[0])),
p=a_dist[0])

                        if (self.name == 'worker_0'):
                          self.env.render()
                        # step
                        s1, r, d, info = self.env.step(a)
                        # normalize reward
                        r = r/100.0

                        if d == False:
                            episode_frames.append(s1)
                        else:
                            s1 = s
                        # collect experience in buffer
                        episode_buffer.append([s,a,r,s1,d,v[0,0]])
```

```
                episode_values.append(v[0,0])

                episode_reward += r
                s = s1
                total_steps += 1
                episode_step_count += 1
```

If we have 25 entries in the buffer, it's time for an update. First, the value is computed and stored in v1, which is then passed to the train() function, which will output the three loss values: value, policy, and entropy. After this, the episode_buffer is reset. If the episode has terminated, we break from the loop. Finally, we print the episode count and reward on the screen. Note that we have used 25 entries as the time to do the update. Feel free to vary this and see how the training is affected by this hyperparameter:

```
# if buffer has 25 entries, time for an update
if len(episode_buffer) == 25 and d != True and episode_step_count !=
max_episode_steps - 1:
    v1 = sess.run(self.local_AC.value,
feed_dict={self.local_AC.inputs:[s]})[0,0]
    value_loss, policy_loss, entropy =
self.train(episode_buffer,sess,gamma,v1)
    episode_buffer = []
    sess.run(self.update_local_ops)

# idiot check to ensure we did not miss update for some unforseen reason
if (len(episode_buffer) > 30):
    print(self.name, "buffer full ", len(episode_buffer))
    sys.exit()

if d == True:
    break

print("episode: ", episode_count, "| worker: ", self.name, "| episode
reward: ", episode_reward, "| step count: ", episode_step_count)
```

After exiting the episode loop, we use the remaining samples in the buffer to train the network. worker _0 contains the global or master network, which we can save by using saver.save. We can also call the self.increment operation to increment the global episode count by one:

```
# Update the network using the episode buffer at the end of the episode
if len(episode_buffer) != 0:
    value_loss, policy_loss, entropy =
self.train(episode_buffer,sess,gamma,0.0)
print("loss: ", self.name, value_loss, policy_loss, entropy)

# write to file for worker_0
```

```
if (self.name == 'worker_0'):
    with open("performance.txt", "a") as myfile:
        myfile.write(str(episode_count) + " " + str(episode_reward) + " " +
str(episode_step_count) + "\n")

# save model params for worker_0
if (episode_count % 25 == 0 and self.name == 'worker_0' and episode_count
!= 0):
        saver.save(sess,self.model_path+'/model-
'+str(episode_count)+'.cptk')
print ("Saved Model")
if self.name == 'worker_0':
    sess.run(self.increment)
episode_count += 1
```

That's it for `a3c.py`.

Coding utils.py

Last but not least, we will code the `utility` functions in `utils.py`. We will import the necessary packages, and we'll also define the `update_target_graph()` function that we used earlier. It takes the scope of the source and destination parameters as arguments, and it copies the parameters from the source to the destination:

```
import numpy as np
import tensorflow as tf
from random import choice



# copy model params
def update_target_graph(from_scope,to_scope):
    from_params = tf.get_collection(tf.GraphKeys.TRAINABLE_VARIABLES,
from_scope)
    to_params = tf.get_collection(tf.GraphKeys.TRAINABLE_VARIABLES,
to_scope)

    copy_ops = []
    for from_param,to_param in zip(from_params,to_params):
        copy_ops.append(to_param.assign(from_param))
    return copy_ops
```

The other utility function that we need is the `discount()` function. It runs the input list, x backwards, and sums them with a weight of `gamma`, which is the discount factor. The discounted value is then returned from the function:

```
# Discounting function used to calculate discounted returns.
def discount(x, gamma):
    dsr = np.zeros_like(x,dtype=np.float32)
    running_sum = 0.0
    for i in reversed(range(0, len(x))):
        running_sum = gamma * running_sum + x[i]
        dsr[i] = running_sum
    return dsr
```

Training on CartPole

The code for `cartpole.py` can be run using the following command:

python cartpole.py

The code stores the episode rewards in the `performance.txt` file. A plot of the episode rewards during training is shown in the following screenshot:

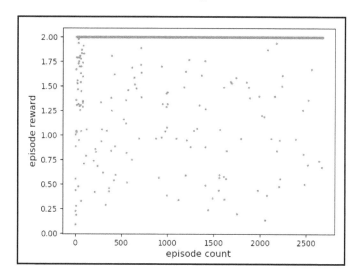

Figure 1: Episode rewards for CartPole, which was trained using A3C

Note that since we have shaped the reward, the episode reward that you can see in the preceding screenshot is different from the values that are typically reported by other researchers in papers and/or blogs.

The A3C algorithm applied to LunarLander

We will extend the same code to train an agent on the LunarLander problem, which is harder than CartPole. Most of the code is the same as before, so we will only describe the changes that need to be made to the preceding code. First, the reward shaping is different for the LunarLander problem. So, we will include a function called `reward_shaping()` in the `a3c.py` file. It will check if the lander has crashed on the lunar surface; if so, the episode will be terminated and there will be a -1.0 penalty. If the lander is not moving, the episode will be terminated and a -0.5 penalty will be paid:

```
def reward_shaping(r, s, s1):
    # check if y-coord < 0; implies lander crashed
    if (s1[1] < 0.0):
      print('-----lander crashed!----- ')
      d = True
      r -= 1.0

    # check if lander is stuck
    xx = s[0] - s1[0]
    yy = s[1] - s1[1]
    dist = np.sqrt(xx*xx + yy*yy)
    if (dist < 1.0e-4):
      print('-----lander stuck!----- ')
      d = True
      r -= 0.5
    return r, d
```

We will call this function after `env.step()`:

```
# reward shaping for lunar lander
r, d = reward_shaping(r, s, s1)
```

Coding lunar.py

The `cartpole.py` file from the previous exercise has been renamed to `lunar.py`. The changes that have been made are as follows. First, we set the maximum number of time steps per episode to `1000` for LunarLander, the discount factor to `gamma = 0.999`, and the state and action dimensions to `8` and `4`, respectively:

```
max_episode_steps = 1000
gamma = 0.999
s_size = 8
a_size = 4
```

The environment is set to `LunarLander-v2`:

```
env = gym.make('LunarLander-v2')
```

That's it for the code changes for training A3C on LunarLander.

Training on LunarLander

You can start the training by using the following command:

```
python lunar.py
```

This will train the agent and store the episode rewards in the `performance.txt` file, which we can plot as follows:

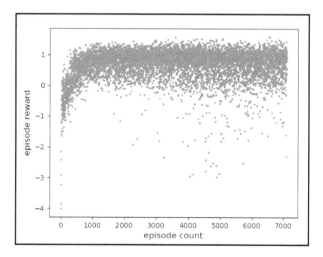

Figure 2: Episode rewards for LunarLander using A3C

As you can see, the agent has learned to land the spacecraft on the lunar surface. Happy landings! Again, note that the episode reward is different from the values that have been reported in papers and blogs by other RL practitioners, since we have scaled the rewards.

The A2C algorithm

The difference between A2C and A3C is that A2C performs synchronous updates. Here, all the workers will wait until they have completed the collection of experiences and computed the gradients. Only after this are the global (or master) network's parameters updated. This is different from A3C, where the update is performed asynchronously, that is, where the worker threads do not wait for the others to finish. A2C is easier to code than A3C, but that is not undertaken here. If you are interested in this, you are encouraged to take the preceding A3C code and convert it to A2C, after which the performance of both algorithms can be compared.

Summary

In this chapter, we introduced the A3C algorithm, which is an on-policy algorithm that's applicable to both discrete and continuous action problems. You saw how three different loss terms are combined into one and optimized. Python's threading library is useful for running multiple threads, with a copy of the policy network in each thread. These different workers compute the policy gradients and pass them on to the master to update the neural network parameters. We applied A3C to train agents for the CartPole and the LunarLander problems, and the agents learned them very well. A3C is a very robust algorithm and does not require a replay buffer, although it does require a local buffer for collecting a small number of experiences, after which it is used to update the networks. Lastly, a synchronous version of the algorithm, called A2C, was also introduced.

This chapter should have really improved your understanding of yet another deep RL algorithm. In the next chapter, we will study the last two RL algorithms in this book, TRPO and PPO.

Questions

1. Is A3C an on-policy or off-policy algorithm?
2. Why is the Shannon entropy term used?
3. What are the problems with using a large number of worker threads?
4. Why is softmax used in the policy neural network?
5. Why do we need an `advantage` function?
6. This is left as an exercise: For the LunarLander problem, repeat the training without reward shaping and see if the agent learns faster/slower than what we saw in this chapter.

Further reading

- *Asynchronous Methods for Deep Reinforcement Learning*, by *Volodymyr Mnih, Adrià Puigdomènech Badia, Mehdi Mirza, Alex Graves, Timothy P. Lillicrap, Tim Harley, David Silver*, and *Koray Kavukcuoglu*, A3C paper from *DeepMind* arXiv:1602.01783: `https://arxiv.org/abs/1602.01783`
- *Deep Reinforcement Learning Hands-On*, by *Maxim Lapan, Packt Publishing*: `https://www.packtpub.com/big-data-and-business-intelligence/deep-reinforcement-learning-hands`

7
Trust Region Policy Optimization and Proximal Policy Optimization

In the last chapter, we saw the use of A3C and A2C, with the former being asynchronous and the latter synchronous. In this chapter, we will see another on-policy **reinforcement learning** (**RL**) algorithm; two algorithms, to be precise, with a lot of similarities in the mathematics, differing, however, in how they are solved. We will be introduced to the algorithm called **Trust Region Policy Optimization** (**TRPO**), which was introduced in 2015 by researchers at OpenAI and the University of California, Berkeley (the latter is incidentally my former employer!). This algorithm, however, is difficult to solve mathematically, as it involves the conjugate gradient algorithm, which is relatively difficult to solve; note that first order optimization methods, such as the well established Adam and **Stochastic Gradient Descent** (**SGD**), cannot be used to solve the TRPO equations. We will then see how solving the policy optimization equations can be combined into one, to result in the **Proximal Policy Optimization** (**PPO**) algorithm, and first order optimization algorithms such as Adam or SGD can be used.

The following topics will be covered in this chapter:

- Learning TRPO
- Learning PPO
- Using PPO to solve the MountainCar problem
- Evaluating the performance

Technical requirements

To successfully complete this chapter, the following software are required:

- Python (2 and above)
- NumPy
- TensorFlow (version 1.4 or higher)

Learning TRPO

TRPO is a very popular on-policy algorithm from OpenAI and the University of California, Berkeley, and was introduced in 2015. There are many flavors of TRPO, but we will learn about the vanilla TRPO version from the paper *Trust Region Policy Optimization, by John Schulman, Sergey Levine, Philipp Moritz, Michael I. Jordan, and Pieter Abbeel, arXiv:1502.05477*: `https://arxiv.org/abs/1502.05477`.

TRPO involves solving a policy optimization equation subject to an additional constraint on the size of the policy update. We will see these equations now.

TRPO equations

TRPO involves the maximization of the expectation of the ratio of the current policy distribution, π_θ, to the old policy distribution, π_θ^{old} (that is, at an earlier time step), multiplied by the advantage function, A_t, subject to an additional constraint that the expectation of the **Kullback-Leibler** (**KL**) divergence of the old and new policy distributions is bounded by a user-specified value, δ:

$$\text{maximize } E\left[\frac{\pi_\theta(a_t|s_t)}{\pi_\theta^{old}(a_t|s_t)}A_t\right]$$
$$\text{subject to } E\left[KL(\pi_\theta^{old}(\cdot|s_t), \pi_\theta(\cdot|s_t))\right] \leq \delta$$

The first equation here is the policy objective, and the second equation is an additional constraint that ensures that the policy update is gradual and does not make large policy updates that can take the policy to regions that are very far away in parameter space.

Since we have two equations that need to be jointly optimized, first-order optimization algorithms, such as Adam and SGD, will not work. Instead, the equations are solved using the conjugate gradient algorithm, making a linear approximation to the first equation, and a quadratic approximation to the second equation. This, however, is mathematically involved, and so we do not present it here in this book. Instead, we will proceed to the PPO algorithm, which is relatively easier to code.

Learning PPO

PPO is an extension to TRPO, and was introduced in 2017 by researchers at OpenAI. PPO is also an on-policy algorithm, and can be applied to discrete action problems as well as continuous actions. It uses the same ratio of policy distributions as in TRPO, but does not use the KL divergence constraint. Specifically, PPO uses three loss functions that are combined into one. We will now see the three loss functions.

PPO loss functions

The first of the three loss functions involved in PPO is called the clipped surrogate objective. Let $r_t(\theta)$ denote the ratio of the new to old policy probability distributions:

$$r_t(\theta) = \frac{\pi_\theta(a_t|s_t)}{\pi_\theta^{old}(a_t|s_t)}$$

The clipped surrogate objective is given by the following equation, where A_t is the advantage function and ε is a hyper parameter; typically, $\varepsilon = 0.1$ or 0.2 is used:

$$L^{clip}(\theta) = E\left[\min(r_t A_t, \mathrm{clip}(r_t, 1 - \epsilon, 1 + \epsilon)A_t)\right]$$

The `clip()` function bounds the ratio between *1-ε* and *1+ε*, thus keeping the ratio bounded within the range. The `min()` function is the minimum function to ensure that the final objective is a lower bound on the unclipped objective.

The second loss function is the L2 norm of the state value function:

$$L^V(\theta) = E\left[\left(V(s_t) - V^{target}\right)^2\right]$$

The third loss is the Shannon entropy of the policy distribution, which comes from information theory:

$$L^{\text{entropy}}(\theta) = E\left[-\log \pi_\theta(s_t)\right]$$

We will now combine the three losses. Note that we need to maximize L^{clip} and $L^{entropy}$, but minimize L^V. So, we define our total PPO loss function as in the following equation, where c_1 and c_2 are positive constants used to scale the terms:

$$L^{\text{PPO}} = L^{\text{clip}} - c_1 L^V + c_2 L^{\text{entropy}}$$

Note that, if we share the neural network parameters between the policy and the value networks, then the preceding L^{PPO} loss function alone can be maximized. On the other hand, if we have separate neural networks for the policy and the value, then we can have separate loss functions as in the following equation, where L^{policy} is maximized and L^{value} is minimized:

$$L^{\text{policy}} = L^{\text{clip}} + c2 L^{\text{entropy}}$$
$$L^{\text{value}} = L^V$$

Notice that the constant c_1 is not required in this latter setting, where we have separate neural networks for the policy and the value. The neural network parameters are updated over multiple iteration steps over a batch of data points, where the number of update steps are specified by the user as hyper parameters.

Using PPO to solve the MountainCar problem

We will solve the MountainCar problem using PPO. MountainCar involves a car trapped in the valley of a mountain. It has to apply throttle to accelerate against gravity and try to drive out of the valley up steep mountain walls to reach a desired flag point on the top of the mountain. You can see a schematic of the MountainCar problem from OpenAI Gym at https://gym.openai.com/envs/MountainCar-v0/.

This problem is very challenging, as the agent cannot just apply full throttle from the base of the mountain and try to reach the flag point, as the mountain walls are steep and gravity will not allow the car to achieve sufficient enough momentum. The optimal solution is for the car to initially go backward and then step on the throttle to pick up enough momentum to overcome gravity and successfully drive out of the valley. We will see that the RL agent actually learns this trick.

We will code the following two files to solve MountainCar using PPO:

- `class_ppo.py`
- `train_test.py`

Coding the class_ppo.py file

We will now code the `class_ppo.py` file:

1. **Import packages**: First, we will import the required packages as follows:

```
import numpy as np
import gym
import sys
```

2. **Set the neural network initializers**: Then, we will set the neural network parameters (we will use two hidden layers) and the initializers for the weights and biases. As we have also done in past chapters, we will use the Xavier initializer for the weights and a small positive value for the initial values of the biases:

```
nhidden1 = 64
nhidden2 = 64


xavier = tf.contrib.layers.xavier_initializer()
bias_const = tf.constant_initializer(0.05)
rand_unif =
tf.keras.initializers.RandomUniform(minval=-3e-3,maxval=3e-3)
regularizer = tf.contrib.layers.l2_regularizer(scale=0.0
```

3. **Define the PPO class**: The PPO() class is now defined. First, the __init__() constructor is defined using the arguments passed to the class. Here, sess is the TensorFlow session; S_DIM and A_DIM are the state and action dimensions, respectively; A_LR and C_LR are the learning rates for the actor and the critic, respectively; A_UPDATE_STEPS and C_UPDATE_STEPS are the number of update steps used for the actor and the critic; CLIP_METHOD stores the epsilon value:

```
class PPO(object):

    def __init__(self, sess, S_DIM, A_DIM, A_LR, C_LR,
A_UPDATE_STEPS, C_UPDATE_STEPS, CLIP_METHOD):
        self.sess = sess
        self.S_DIM = S_DIM
        self.A_DIM = A_DIM
        self.A_LR = A_LR
        self.C_LR = C_LR
        self.A_UPDATE_STEPS = A_UPDATE_STEPS
        self.C_UPDATE_STEPS = C_UPDATE_STEPS
        self.CLIP_METHOD = CLIP_METHOD
```

4. **Define TensorFlow placeholders**: We will next need to define the TensorFlow placeholders: tfs for the state, tfdc_r for the discounted rewards, tfa for the actions, and tfadv for the advantage function:

```
# tf placeholders
self.tfs = tf.placeholder(tf.float32, [None, self.S_DIM], 'state')
self.tfdc_r = tf.placeholder(tf.float32, [None, 1], 'discounted_r')
self.tfa = tf.placeholder(tf.float32, [None, self.A_DIM], 'action')
self.tfadv = tf.placeholder(tf.float32, [None, 1], 'advantage')
```

5. **Define the critic**: The critic neural network is defined next. We use the state (s_t) placeholder, self.tfs, as input to the neural network. Two hidden layers are used with the nhidden1 and nhidden2 number of neurons and the relu activation function (both nhidden1 and nhidden2 were set to 64 previously). The output layer has one neuron that will output the state value function $V(s_t)$, and so no activation function is used for the output. We then compute the advantage function as the difference between the discounted cumulative rewards, which is stored in the self.tfdc_r placeholder and the self.v output that we just computed. The critic loss is computed as an L2 norm and the critic is trained using the Adam optimizer with the objective to minimize this L2 loss.

Note that this loss is the same as L^{value} mentioned earlier in this chapter in the theory section:

```
# critic
with tf.variable_scope('critic'):
    l1 = tf.layers.dense(self.tfs, nhidden1, activation=None,
kernel_initializer=xavier, bias_initializer=bias_const,
kernel_regularizer=regularizer)
    l1 = tf.nn.relu(l1)
    l2 = tf.layers.dense(l1, nhidden2, activation=None,
kernel_initializer=xavier, bias_initializer=bias_const,
kernel_regularizer=regularizer)
    l2 = tf.nn.relu(l2)
    self.v = tf.layers.dense(l2, 1, activation=None,
kernel_initializer=rand_unif, bias_initializer=bias_const)
    self.advantage = self.tfdc_r - self.v
    self.closs = tf.reduce_mean(tf.square(self.advantage))
    self.ctrain_op =
tf.train.AdamOptimizer(self.C_LR).minimize(self.closs)
```

6. **Call the _build_anet function**: We define the actor using a `_build_anet()` function that will soon be specified. Specifically, the policy distribution and the list of model parameters are output from this function. We call this function once for the current policy and again for the older policy. The mean and standard deviation can be obtained from `self.pi` by calling the `mean()` and `stddev()` functions, respectively:

```
# actor
self.pi, self.pi_params = self._build_anet('pi', trainable=True)
self.oldpi, self.oldpi_params = self._build_anet('oldpi',
trainable=False)

self.pi_mean = self.pi.mean()
self.pi_sigma = self.pi.stddev()
```

7. **Sample actions**: From the policy distribution, `self.pi`, we can also sample actions using the `sample()` function that is part of TensorFlow distributions:

```
with tf.variable_scope('sample_action'):
    self.sample_op = tf.squeeze(self.pi.sample(1), axis=0)
```

8. **Update older policy parameters**: The older policy network parameters can be updated using the new policy values simply by assigning the values from the latter to the former, using TensorFlow's `assign()` function. Note that the new policy is optimized – the older policy is simply a copy of the current policy, albeit from one update cycle earlier:

```
with tf.variable_scope('update_oldpi'):
    self.update_oldpi_op = [oldp.assign(p) for p, oldp in
zip(self.pi_params, self.oldpi_params)]
```

9. **Compute policy distribution ratio**: The policy distribution ratio is computed at the `self.tfa` action, and is stored in `self.ratio`. Note that, exponentially, the difference of logarithms of the distributions is the same as the ratio of the distributions. This ratio is then clipped to bound it between $1-\varepsilon$ and $1+\varepsilon$, as explained earlier in the theory:

```
with tf.variable_scope('loss'):
    self.ratio = tf.exp(self.pi.log_prob(self.tfa) -
self.oldpi.log_prob(self.tfa))
    self.clipped_ratio = tf.clip_by_value(self.ratio, 1.-
self.CLIP_METHOD['epsilon'], 1.+self.CLIP_METHOD['epsilon'])
```

10. **Compute losses**: The total loss for the policy, as mentioned previously, involves three losses that are combined when the policy and value neural networks share weights. However, since we consider the other setting mentioned in the theory earlier in this chapter, where we have separate neural networks for the policy and the value, we will have two losses for the policy optimization. The first is the minimum of the product of the unclipped ratio and the advantage function and its clipped analogue—this is stored in `self.aloss`. The second loss is the Shannon entropy, which is the product of the policy distribution and its logarithm, summed over, and a minus sign included. This term is scaled with the hyper parameter, $c_1 = 0.01$, and subtracted from the loss. For the time being, the entropy loss term is set to zero, as it also is in the PPO paper. We can consider including this entropy loss later to see if it makes any difference in the learning of the policy. We use the Adam optimizer. Note that we need to maximize the original policy loss mentioned in the theory earlier in this chapter, but the Adam optimizer has the `minimize()` function, so we have included a minus sign in `self.aloss` (see the first line of the following code), as maximizing a loss is the same as minimizing the negative of it:

```
self.aloss = -tf.reduce_mean(tf.minimum(self.ratio*self.tfadv,
self.clipped_ratio*self.tfadv))

# entropy
```

```
entropy = -tf.reduce_sum(self.pi.prob(self.tfa) *
tf.log(tf.clip_by_value(self.pi.prob(self.tfa),1e-10,1.0)),axis=1)
entropy = tf.reduce_mean(entropy,axis=0)
self.aloss -= 0.0 #0.01 * entropy


with tf.variable_scope('atrain'):
    self.atrain_op =
tf.train.AdamOptimizer(self.A_LR).minimize(self.aloss)
```

11. **Define the update function**: The `update()` function is defined next, which takes the `s` state, the `a` action, and the `r` reward as arguments. It involves running a TensorFlow session on updating the old policy network parameters by calling the TensorFlow `self.update_oldpi_op` operation. Then, the advantage is computed, which, along with the state and action, is used to update the `A_UPDATE_STEPS` actor number of iterations. Then, the critic is updated by the `C_UPDATE_STEPS` number of iterations by running a TensorFlow session on the critic train operation:

```
def update(self, s, a, r):
    self.sess.run(self.update_oldpi_op)
    adv = self.sess.run(self.advantage, {self.tfs: s, self.tfdc_r:
r})


    # update actor
    for _ in range(self.A_UPDATE_STEPS):
        self.sess.run(self.atrain_op, feed_dict={self.tfs: s,
self.tfa: a, self.tfadv: adv})
    # update critic
    for _ in range(self.C_UPDATE_STEPS):
        self.sess.run(self.ctrain_op, {self.tfs: s, self.tfdc_r:
r})
```

12. **Define the _build_anet function**: We will next define the `_build_anet()` function that was used earlier. It will compute the policy distribution, which is treated as a Gaussian (that is, normal). It takes the `self.tfs` state placeholder as input, has two hidden layers with the `nhidden1` and `nhidden2` neurons, and uses the `relu` activation function. This is then sent to two output layers with the `A_DIM` action dimension number of outputs, with one representing the mean, `mu`, and the other the standard deviation, `sigma`.

Note that the mean of the actions are bounded, and so the `tanh` activation function is used, including a small clipping to avoid edge values; for sigma, the `softplus` activation function is used, shifted by `0.1` to avoid zero sigma values. Once we have the mean and standard deviations for the actions, TensorFlow distributions' `Normal` is used to treat the policy as a Gaussian distribution. We can also call `tf.get_collection()` to obtain the model parameters, and the `Normal` distribution and the model parameters are returned from the function:

```
def _build_anet(self, name, trainable):
    with tf.variable_scope(name):
        l1 = tf.layers.dense(self.tfs, nhidden1, activation=None,
trainable=trainable, kernel_initializer=xavier,
bias_initializer=bias_const, kernel_regularizer=regularizer)
        l1 = tf.nn.relu(l1)
        l2 = tf.layers.dense(l1, nhidden2, activation=None,
trainable=trainable, kernel_initializer=xavier,
bias_initializer=bias_const, kernel_regularizer=regularizer)
        l2 = tf.nn.relu(l2)
        mu = tf.layers.dense(l2, self.A_DIM, activation=tf.nn.tanh,
trainable=trainable, kernel_initializer=rand_unif,
bias_initializer=bias_const)

        small = tf.constant(1e-6)
        mu = tf.clip_by_value(mu,-1.0+small,1.0-small)

        sigma = tf.layers.dense(l2, self.A_DIM, activation=None,
trainable=trainable, kernel_initializer=rand_unif,
bias_initializer=bias_const)
        sigma = tf.nn.softplus(sigma) + 0.1

        norm_dist = tf.distributions.Normal(loc=mu, scale=sigma)
    params = tf.get_collection(tf.GraphKeys.GLOBAL_VARIABLES,
scope=name)
        return norm_dist, params
```

13. **Define the choose_action function**: We also define a `choose_action()` function to sample from the policy to obtain actions:

```
def choose_action(self, s):
    s = s[np.newaxis, :]
    a = self.sess.run(self.sample_op, {self.tfs: s})
    return a[0]
```

14. **Define the get_v function**: Finally, we also define a `get_v()` function to return the state value by running a TensorFlow session on `self.v`:

```
def get_v(self, s):
    if s.ndim < 2: s = s[np.newaxis, :]
    vv = self.sess.run(self.v, {self.tfs: s})
    return vv[0,0]
```

That concludes `class_ppo.py`. We will now code `train_test.py`.

Coding train_test.py file

We will now code the `train_test.py` file.

1. **Importing the packages:** First, we import the required packages:

```
import tensorflow as tf
import numpy as np
import matplotlib.pyplot as plt
import gym
import sys
import time

from class_ppo import *
```

2. **Define function:** We then define a function for reward shaping that will give out some extra bonus rewards and penalties for good and bad performance, respectively. We do this for encouraging the car to go higher towards the side of the flag which is on the mountain top, without which the learning will be slow:

```
def reward_shaping(s_):

    r = 0.0

    if s_[0] > -0.4:
        r += 5.0*(s_[0] + 0.4)
    if s_[0] > 0.1:
        r += 100.0*s_[0]
    if s_[0] < -0.7:
        r += 5.0*(-0.7 - s_[0])
    if s_[0] < 0.3 and np.abs(s_[1]) > 0.02:
        r += 4000.0*(np.abs(s_[1]) - 0.02)

    return r
```

3. We next choose `MountainCarContinuous` as the environment. The total number of episodes we will train the agent for is `EP_MAX`, and we set this to `1000`. The `GAMMA` discount factor is set to `0.9` and the learning rates to `2e-4`. We use a batch size of `32` and perform `10` update steps per cycle. The state and action dimensions are obtained and stored in `S_DIM` and `A_DIM`, respectively. For the PPO `clip` parameter, `epsilon`, we use a value of `0.1`. `train_test` is set to `0` for training the agent and `1` for testing:

```python
env = gym.make('MountainCarContinuous-v0')


EP_MAX = 1000
GAMMA = 0.9

A_LR = 2e-4
C_LR = 2e-4

BATCH = 32
A_UPDATE_STEPS = 10
C_UPDATE_STEPS = 10

S_DIM = env.observation_space.shape[0]
A_DIM = env.action_space.shape[0]

print("S_DIM: ", S_DIM, "| A_DIM: ", A_DIM)

CLIP_METHOD = dict(name='clip', epsilon=0.1)

# train_test = 0 for train; =1 for test
train_test = 0

# irestart = 0 for fresh restart; =1 for restart from ckpt file
irestart = 0

iter_num = 0

if (irestart == 0):
    iter_num = 0
```

4. We create a TensorFlow session and call it `sess`. An instance of the `PPO` class is created, called `ppo`. We also create a TensorFlow saver. Then, if we are training from scratch, we initialize all the model parameters by calling `tf.global_variables_initializer()`, or, if we are continuing the training from a saved agent or testing, then we restore from the `ckpt/model` path:

```
sess = tf.Session()

ppo = PPO(sess, S_DIM, A_DIM, A_LR, C_LR, A_UPDATE_STEPS,
C_UPDATE_STEPS, CLIP_METHOD)

saver = tf.train.Saver()


if (train_test == 0 and irestart == 0):
  sess.run(tf.global_variables_initializer())
else:
  saver.restore(sess, "ckpt/model")
```

5. The main `for loop` over episodes is then defined. Inside it, we reset the environment and also set buffers to empty lists. The terminal Boolean, `done`, and the number of time steps, `t`, are also initialized:

```
for ep in range(iter_num, EP_MAX):

    print("-"*70)
    s = env.reset()

    buffer_s, buffer_a, buffer_r = [], [], []
    ep_r = 0

    max_pos = -1.0
    max_speed = 0.0
    done = False
    t = 0
```

Inside the outer loop, we have the inner `while` loop over time steps. This problem involves short time steps during which the car may not significantly move, and so we use sticky actions where actions are sampled from the policy only once every 8 time steps. The `choose_action()` function in the PPO class will sample the actions for a given state. A small Gaussian noise is added to the actions to explore, and are clipped in the −1.0 to 1.0 range, as required for the `MountainCarContinuous` environment. The action is then fed into the environment's `step()` function, which will output the next `s_` state, `r` reward, and the terminal `done` Boolean. The `reward_shaping()` function is called to shape rewards. To track how far the agent is pushing its limits, we also compute its maximum position and speed in `max_pos` and `max_speed`, respectively:

```
while not done:
    env.render()

    # sticky actions
    #if (t == 0 or np.random.uniform() < 0.125):
    if (t % 8 ==0):
      a = ppo.choose_action(s)

    # small noise for exploration
    a += 0.1 * np.random.randn()

    # clip
    a = np.clip(a, -1.0, 1.0)

    # take step
    s_, r, done, _ = env.step(a)
    if s_[0] > 0.4:
        print("nearing flag: ", s_, a)

    if s_[0] > 0.45:
      print("reached flag on mountain! ", s_, a)
      if done == False:
        print("something wrong! ", s_, done, r, a)
        sys.exit()

    # reward shaping
    if train_test == 0:
      r += reward_shaping(s_)

    if s_[0] > max_pos:
        max_pos = s_[0]
    if s_[1] > max_speed:
        max_speed = s_[1]
```

6. If we are in training mode, the state, action, and reward are appended to the buffer. The new state is set to the current state and we proceed to the next time step if the episode has not already terminated. The `ep_r` episode total rewards and the `t` time step count are also updated:

```
if (train_test == 0):
    buffer_s.append(s)
    buffer_a.append(a)
    buffer_r.append(r)

    s = s_
    ep_r += r
    t += 1
```

If we are in the training mode, if the number of samples is equal to a batch, or if the episode has terminated, we will train the neural networks. For this, the state value for the new state is first obtained using `ppo.get_v`. Then, we compute the discounted rewards. The buffer lists are also converted to NumPy arrays, and the buffer lists are reset to empty lists. These `bs`, `ba`, and `br` NumPy arrays are then used to update the `ppo` object's actor and critic networks:

```
if (train_test == 0):
    if (t+1) % BATCH == 0 or done == True:
        v_s_ = ppo.get_v(s_)
        discounted_r = []
        for r in buffer_r[::-1]:
            v_s_ = r + GAMMA * v_s_
            discounted_r.append(v_s_)
            discounted_r.reverse()

        bs = np.array(np.vstack(buffer_s))
        ba = np.array(np.vstack(buffer_a))
        br = np.array(discounted_r)[:, np.newaxis]

        buffer_s, buffer_a, buffer_r = [], [], []
        ppo.update(bs, ba, br)
```

7. If we are in testing mode, Python is paused briefly for better visualization. If the episode has terminated, the `while` loop is exited with a `break` statement. Then, we print the maximum position and speed values on the screen, as well as write them, along with the episode rewards, to a file called `performance.txt`. Once every 10 episodes, the model is also saved by calling `saver.save`:

```
if (train_test == 1):
    time.sleep(0.1)
```

```
        if (done == True):
            print("values at done: ", s_, a)
            break

        print("episode: ", ep, "| episode reward: ", round(ep_r,4), "| time
    steps: ", t)
        print("max_pos: ", max_pos, "| max_speed:", max_speed)

        if (train_test == 0):
            with open("performance.txt", "a") as myfile:
                myfile.write(str(ep) + " " + str(round(ep_r,4)) + " " +
    str(round(max_pos,4)) + " " + str(round(max_speed,4)) + "\n")

        if (train_test == 0 and ep%10 == 0):
            saver.save(sess, "ckpt/model")
```

This concludes the coding of PPO. We will next evaluate its performance on MountainCarContinuous.

Evaluating the performance

The PPO agent is trained by the following command:

```
python train_test.py
```

Once the training is complete, we can test the agent by setting the following:

```
train_test = 1
```

Then, we will repeat `python train_test.py` again. On visualizing the agent, we can observe that the car first moves backward to climb the left mountain. Then it goes full throttle and picks up enough momentum to drive past the steep slope of the right mountain with the flag on top. So, the PPO agent has learned to drive out of the mountain valley successfully.

Full throttle

Note that we had to navigate backward first and then step on the throttle in order to have sufficient momentum to escape gravity and successfully drive out of the mountain valley. What if we had just stepped on the throttle right from the first step – would the car still be able to escape? Let's check by coding and running `mountaincar_full_throttle.py`.

We will now set the action to `1.0`, that is, full throttle:

```
import sys
import numpy as np
import gym

env = gym.make('MountainCarContinuous-v0')


for _ in range(100):
  s = env.reset()
  done = False

  max_pos = -1.0
  max_speed = 0.0
  ep_reward = 0.0

  while not done:
    env.render()
    a = [1.0] # step on throttle
    s_, r, done, _ = env.step(a)

    if s_[0] > max_pos: max_pos = s_[0]
    if s_[1] > max_speed: max_speed = s_[1]
    ep_reward += r

  print("ep_reward: ", ep_reward, "| max_pos: ", max_pos, "| max_speed: ",
max_speed)
```

As is evident from the video generated during the training, the car is unable to escape the inexorable pull of gravity, and remains stuck at the base of the mountain valley.

Random throttle

What if we try random throttle values? We will code `mountaincar_random_throttle.py` with random actions in the `-1.0` to `1.0` range:

```
import sys
import numpy as np
import gym

env = gym.make('MountainCarContinuous-v0')


for _ in range(100):
  s = env.reset()
```

```
        done = False

    max_pos = -1.0
    max_speed = 0.0
    ep_reward = 0.0

    while not done:
      env.render()
      a = [-1.0 + 2.0*np.random.uniform()]
      s_, r, done, _ = env.step(a)

      if s_[0] > max_pos: max_pos = s_[0]
      if s_[1] > max_speed: max_speed = s_[1]
      ep_reward += r

    print("ep_reward: ", ep_reward, "| max_pos: ", max_pos, "| max_speed: ",
max_speed)
```

Here too, the car fails to escape gravity and remains stuck at the base. So, the RL agent is required to figure out that the optimum policy here is to first go backward, and then step on the throttle to escape gravity and reach the flag on the mountain top.

This concludes our MountainCar exercise with PPO.

Summary

In this chapter, we were introduced to the TRPO and PPO RL algorithms. TRPO involves two equations that need to be solved, with the first equation being the policy objective and the second equation being a constraint on how much we can update. TRPO requires second-order optimization methods, such as conjugate gradient. To simplify this, the PPO algorithm was introduced, where the policy ratio is clipped within a certain user-specified range so as to keep the update gradual. In addition, we also saw the use of data samples collected from experience to update the actor and the critic for multiple iteration steps. We trained the PPO agent on the MountainCar problem, which is a challenging problem, as the actor must first drive the car backward up the left mountain, and then accelerate to gain sufficient momentum to overcome gravity and reach the flag point on the right mountain. We also saw that a full throttle policy or a random policy will not help the agent reach its goal.

With this chapter, we have looked at several RL algorithms. In the next chapter, we will apply DDPG and PPO to train an agent to drive a car autonomously.

Questions

1. Can we apply Adam or SGD optimization in TRPO?
2. What is the role of the entropy term in the policy optimization?
3. Why do we clip the policy ratio? What will happen if the clipping parameter epsilon is large?
4. Why do we use the `tanh` activation function for `mu` and `softplus` for sigma? Can we use the `tanh` activation function for sigma?
5. Does reward shaping always help in the training?
6. Do we need reward shaping when we test an already trained agent?

Further reading

- *Trust Region Policy Optimization, John Schulman, Sergey Levine, Philipp Moritz, Michael I. Jordan, Pieter Abbeel,* arXiv:1502.05477 (TRPO paper): `https://arxiv.org/abs/1502.05477`

- *Proximal Policy Optimization Algorithms, John Schulman, Filip Wolski, Prafulla Dhariwal, Alec Radford, Oleg Klimov,* arXiv:1707.06347 (PPO paper): `https://arxiv.org/abs/1707.06347`

- *Deep Reinforcement Learning Hands-On, Maxim Lapan, Packt Publishing*: `https://www.packtpub.com/big-data-and-business-intelligence/deep-reinforcement-learning-hands`

8
Deep RL Applied to Autonomous Driving

Autonomous driving is one of the hottest technological revolutions in development as of the time of writing this. It will dramatically alter how humanity looks at transportation in general, and will drastically reduce travel costs as well as increase safety. Several state-of-the-art algorithms are used by the autonomous vehicle development community to this end. These include, but are not limited to, perception, localization, path planning, and control. Perception deals with the identification of the environment around an autonomous vehicle—pedestrians, cars, bicycles, and so on. Localization involves the identification of the exact location—or pose to be more precise—of the vehicle in a precomputed map of the environment. Path planning, as the name implies, is the process of planning the path of the autonomous vehicle, both in the long term (say, from point *A* to point *B*) as well as the shorter term (say, the next 5 seconds). Control is the actual execution of the desired path, including evasive maneuvers. In particular, **reinforcement learning** (**RL**) is widely used in the path planning and control of the autonomous vehicle, both for urban as well as highway driving.

In this chapter, we will use the **The Open Racing Car Simulator** (**TORCS**) simulator to train an RL agent to learn to successfully drive on a racetrack. While the CARLA simulator is more robust and has realistic rendering, TORCS is easier to use and so is a good first option. The interested reader is encouraged to try out training RL agents on the CARLA simulator after completing this book.

The following topics will be covered in this chapter:

- Learning to use TORCS
- Training a **Deep Deterministic Policy Gradient** (**DDPG**) agent to learn to drive
- Training a **Proximal Policy Optimization** (**PPO**) agent

Technical requirements

To complete this chapter, we will need the following:

- Python (version 2 or 3)
- NumPy
- Matplotlib
- TensorFlow (version 1.4 or higher)
- TORCS racing car simulator

Car driving simulators

Applying RL in autonomous driving necessitates the use of robust car-driving simulators, as the RL agent cannot be trained on the road directly. To this end, several open source car-driving simulators have been developed by the research community, with each having its own pros and cons. Some of the open source car driving simulators are:

- CARLA
  - http://vladlen.info/papers/carla.pdf
  - Developed at Intel labs
  - Suited to urban driving
- TORCS
  - http://torcs.sourceforge.net/
  - Racing car
- DeepTraffic
  - https://selfdrivingcars.mit.edu/deeptraffic/
  - Developed at MIT
  - Suited to highway driving

Learning to use TORCS

We will first learn how to use the TORCS racing car simulator, which is an open source simulator. You can obtain the download instructions from `http://torcs.sourceforge.net/index.php?name=Sectionsop=viewarticleartid=3` but the salient steps are summarized as follows for Linux:

1. Download the `torcs-1.3.7.tar.bz2` file from `https://sourceforge.net/projects/torcs/files/all-in-one/1.3.7/torcs-1.3.7.tar.bz2/download`
2. Unpack the package with `tar xfvj torcs-1.3.7.tar.bz2`
3. Run the following commands:
   - `cd torcs-1.3.7`
   - `./configure`
   - `make`
   - `make install`
   - `make datainstall`
4. The default installation directories are:

   - `/usr/local/bin`: TORCS command (directory should be in your `PATH`)
   - `/usr/local/lib/torcs`: TORCS dynamic `libs` (directory MUST be in your `LD_LIBRARY_PATH` if you don't use the TORCS shell)
   - `/usr/local/share/games/torcs`: TORCS data files

By running the `torcs` command (the default location is `/usr/local/bin/torcs`), you can now see the TORCS simulator open. The desired settings can then be chosen, including the choice of the car, racetrack, and so on. The simulator can also be played as a video game, but we are interested in using it to train an RL agent.

State space

We will next define the state space for TORCS. *Table 2: Description of the available sensors (part II). Ranges are reported with their unit of measure (where defined)* of the *Simulated Car Racing Championship: Competition Software Manual* document at `https://arxiv.org/pdf/1304.1672.pdf` provides a summary of the state parameters that are available for the simulator. We will use the following entries as our state space; the number in brackets identifies the size of the entry:

- `angle`: Angle between the car direction and the track (1)
- `track`: This will give us the end of the track measured every 10 degrees from -90 to +90 degrees; it has 19 real values, counting the end values (19)
- `trackPos`: Distance between the car and the track axis (1)
- `speedX`: Speed of the car in the longitudinal direction (1)
- `speedY`: Speed of the car in the transverse direction (1)
- `speedZ`: Speed of the car in the Z-direction; we don't need this actually, but we retain it for now (1)
- `wheelSpinVel`: The rotational speed of the four wheels of the car (4)
- `rpm`: The car engine's rpm (1)

See the previously mentioned document for a better understanding of the preceding variables, including their permissible ranges. Summing up the number of real valued entries, we note that our state space is a real valued vector of size 1+19+1+1+1+1+4+1 = 29. Our action space is of size *3*: the steering, acceleration, and brake. Steering is in the range *[-1,1]*, and acceleration is in the range *[0,1]*, as is the brake.

Support files

The open source community has also developed two Python files to interface TORCS with Python so that we can call TORCS from Python commands. In addition, to automatically start TORCS, we need another `sh` file. These three files are summarized as follows:

- `gym_torcs.py`
- `snakeoil3_gym.py`
- `autostart.sh`

These files are included in the code files for the chapter (`https://github.com/PacktPublishing/TensorFlow-Reinforcement-Learning-Quick-Start-Guide`), but can also be obtained from a Google search. In lines ~130-160 of `gym_torcs.py`, the reward function is set. You can see the following lines, which convert the raw simulator states to NumPy arrays:

```
# Reward setting Here ####################################
# direction-dependent positive reward
track = np.array(obs['track'])
trackPos = np.array(obs['trackPos'])
sp = np.array(obs['speedX'])
damage = np.array(obs['damage'])
rpm = np.array(obs['rpm'])
```

The reward function is then set as follows. Note that we give rewards for higher longitudinal speed along the track (the cosine of the angle term), and penalize lateral speed (the sine of the angle term). Track position is also penalized. Ideally, if this were zero, we would be at the center of the track, and values of *+1* or *-1* imply that we are at the edges of the track, which is not desired and hence penalized:

```
progress = sp*np.cos(obs['angle']) - np.abs(sp*np.sin(obs['angle'])) - sp *
np.abs(obs['trackPos'])
reward = progress
```

We terminate the episode if the car is out of the track and/or the progress of the agent is stuck using the following code:

```
if (abs(track.any()) > 1 or abs(trackPos) > 1): # Episode is terminated if
the car is out of track
    print("Out of track ")
    reward = -100 #-200
    episode_terminate = True
    client.R.d['meta'] = True

if self.terminal_judge_start < self.time_step: # Episode terminates if the
progress of agent is small
    if progress < self.termination_limit_progress:
        print("No progress", progress)
        reward = -100 # KAUSHIK ADDED THIS
        episode_terminate = True
        client.R.d['meta'] = True
```

We are now ready to train an RL agent to drive a car in TORCS successfully. We will use a DDPG agent first.

Training a DDPG agent to learn to drive

Most of the DDPG code is the same as we saw earlier in Chapter 5, *Deep Deterministic Policy Gradients (DDPG)*; only the differences will be summarized here.

Coding ddpg.py

Our state dimension for TORCS is 29 and the action dimension is 3; these are set in ddpg.py as follows:

```
state_dim = 29
action_dim = 3
action_bound = 1.0
```

Coding AandC.py

The actor and critic file, AandC.py, also needs to be modified. In particular, the create_actor_network in the ActorNetwork class is edited to have two hidden layers with 400 and 300 neurons, respectively. Also, the output consists of three actions: steering, acceleration, and brake. Since steering is in the [-1,1] range, the tanh activation function is used; acceleration and brake are in the [0,1] range, and so the sigmoid activation function is used. We then concat them along axis dimension 1, and this is the output of our actor's policy:

```
    def create_actor_network(self, scope):
        with tf.variable_scope(scope, reuse=tf.AUTO_REUSE):
            state = tf.placeholder(name='a_states', dtype=tf.float32,
shape=[None, self.s_dim])
            net = tf.layers.dense(inputs=state, units=400, activation=None,
kernel_initializer=winit, bias_initializer=binit, name='anet1')
            net = tf.nn.relu(net)

            net = tf.layers.dense(inputs=net, units=300, activation=None,
kernel_initializer=winit, bias_initializer=binit, name='anet2')
            net = tf.nn.relu(net)
            steering = tf.layers.dense(inputs=net, units=1,
activation=tf.nn.tanh, kernel_initializer=rand_unif,
bias_initializer=binit, name='steer')
            acceleration = tf.layers.dense(inputs=net, units=1,
activation=tf.nn.sigmoid, kernel_initializer=rand_unif,
bias_initializer=binit, name='acc')
            brake = tf.layers.dense(inputs=net, units=1,
```

```
activation=tf.nn.sigmoid, kernel_initializer=rand_unif,
bias_initializer=binit, name='brake')
            out = tf.concat([steering, acceleration, brake], axis=1)

            return state, out
```

Likewise, the `CriticNetwork` class `create_critic_network()` function is edited to have two hidden layers for the neural network, with 400 and 300 neurons, respectively. This is shown in the following code:

```
    def create_critic_network(self, scope):
        with tf.variable_scope(scope, reuse=tf.AUTO_REUSE):
            state = tf.placeholder(name='c_states', dtype=tf.float32,
shape=[None, self.s_dim])
            action = tf.placeholder(name='c_action', dtype=tf.float32,
shape=[None, self.a_dim])

            net = tf.concat([state, action],1)

            net = tf.layers.dense(inputs=net, units=400, activation=None,
kernel_initializer=winit, bias_initializer=binit, name='cnet1')
            net = tf.nn.relu(net)

            net = tf.layers.dense(inputs=net, units=300, activation=None,
kernel_initializer=winit, bias_initializer=binit, name='cnet2')
            net = tf.nn.relu(net)

            out = tf.layers.dense(inputs=net, units=1, activation=None,
kernel_initializer=rand_unif, bias_initializer=binit, name='cnet_out')
            return state, action, out
```

The other changes to be made are in `TrainOrTest.py`, which we will look into next.

Coding TrainOrTest.py

Import the TORCS environment from `gym_torcs` so that we can train the RL agent on it:

1. **Import TORCS**: Import the TORCS environment from `gym_torcs` as follows:

   ```
   from gym_torcs import TorcsEnv
   ```

2. **The** `env` **variable**: Create a TORCS environment variable using the following command:

   ```
   # Generate a Torcs environment
   env = TorcsEnv(vision=False, throttle=True, gear_change=False)
   ```

3. **Relaunch TORCS**: Since TORCS is known to have a memory leak error, reset the environment every `100` episodes using `relaunch=True`; otherwise reset without any arguments as follows:

```
if np.mod(i, 100) == 0:
    ob = env.reset(relaunch=True) #relaunch TORCS every N episodes
due to a memory leak error
else:
    ob = env.reset()
```

4. **Stack up state space**: Use the following command to stack up the 29-dimension space:

```
s = np.hstack((ob.angle, ob.track, ob.trackPos, ob.speedX,
ob.speedY, ob.speedZ, ob.wheelSpinVel/100.0, ob.rpm))
```

5. **Number of time steps per episode**: Choose the number of time steps `msteps` to run per episode. For the first `100` episodes, the agent has not learned much, and so you can choose `100` time steps per episode; we gradually increase this linearly for later episodes up to the upper limit of `max_steps`.

This step is not critical and the agent's learning is not dependent on the number of steps we choose per episode. Feel free to experiment with how `msteps` is set.

Choose the number of time steps as follows:

```
msteps = max_steps
if (i < 100):
    msteps = 100
elif (i >=100 and i < 200):
    msteps = 100 + (i-100)*9
else:
    msteps = 1000 + (i-200)*5
    msteps = min(msteps, max_steps)
```

6. **Full throttle**: For the first `10` episodes, we apply full throttle to warm up the neural network parameters. Only after that, do we start using the actor's policy. Note that TORCS typically learns in about ~1,500–2,000 episodes, so the first `10` episodes will not really have much influence later on in the learning. Apply full throttle to warm up the neural network parameters as follows:

```
# first few episodes step on gas!
if (i < 10):
    a[0][0] = 0.0
```

```
a[0][1] = 1.0
a[0][2] = 0.0
```

That's it for the changes that need to be made to the code for the DDPG to play TORCS. The rest of the code is the same as that covered in `Chapter 5`, *Deep Deterministic Policy Gradients (DDPG)*. We can train the agent using the following command:

`python ddpg.py`

Enter `1` for training; `0` is for testing a pretrained agent. Training can take about 2–5 days depending on the speed of the computer used. But this is a fun problem and is worth the effort. The number of steps experienced per episode, as well as the rewards, are stored in `analysis_file.txt`, which we can plot. The number of time steps per episode is plotted as follows:

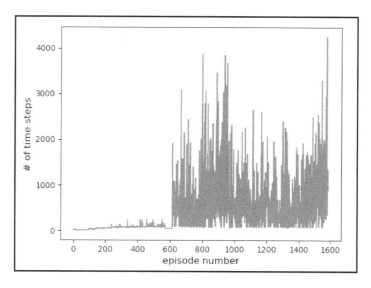

Figure 1: Number of time steps per episode of TORCS (training mode)

We can see that the car has learned to drive reasonably well after ~600 episodes, with more efficient driving after ~1,500 episodes. Approximately ~300 time steps correspond to one lap of the racetrack. Thus, the agent is able to drive more than seven to eight laps without terminating toward the end of the training. For a cool video of the DDPG agent driving, see the following YouTube link: `https://www.youtube.com/watch?v=ajomz08hSIE`.

Training a PPO agent

We saw previously how to train a DDPG agent to drive a car on TORCS. How to use a PPO agent is left as an exercise for the interested reader. This is a nice challenge to complete. The PPO code from `Chapter 7`, *Trust Region Policy Optimization and Proximal Policy Optimization*, can be reused, with the necessary changes made to the TORCS environment. The PPO code for TORCS is also supplied in the code repository (`https://github.com/PacktPublishing/TensorFlow-Reinforcement-Learning-Quick-Start-Guide`), and the interested reader can peruse it. A cool video of a PPO agent driving a car in TORCS is in the following YouTube video at: `https://youtu.be/uE8QaJQ7zDI`

Another challenge for the interested reader is to use **Trust Region Policy Optimization** (**TRPO**) for the TORCS racing car problem. Try this too, if interested! This is one way to master RL algorithms.

Summary

In this chapter, we saw how to apply RL algorithms to train an agent to learn to drive a car autonomously. We installed the TORCS racing-car simulator and also learned how to interface it with Python, so that we can train RL agents. We also did a deep dive into the state space for TORCS and the meaning of each of these terms. The DDPG algorithm was then used to train an agent to learn to drive successfully in TORCS. The video rendering in TORCS is really cool! The trained agent was able to drive more than seven to eight laps around the racetrack successfully. Finally, the use of PPO for the same problem of driving a car autonomously was also explored and left as an exercise for the interested reader; code for this is supplied in the book's repository.

This concludes this chapter as well as the book. Feel free to read upon more material online on the application of RL for autonomous driving and robotics. This is now a very hot area of both academic and industry research, and is well funded, with several job openings in these areas. Wishing you the best!

Questions

1. Why can you not use DQN for the TORCS problem?
2. We used the Xavier weights initializer for the neural network weights. What other weight initializers are you aware of, and how well will the trained agent perform with them?
3. Why is the `abs()` function used in the reward function, and why is it used for the last two terms but not for the first term?
4. How can you ensure smoother driving than what was observed in the video?
5. Why is a replay buffer used in DDPG but not in PPO?

Further reading

- *Continuous control with deep reinforcement learning*, Timothy P. Lillicrap, Jonathan J. Hunt, Alexander Pritzel, Nicolas Heess, Tom Erez, Yuval Tassa, David Silver, Daan Wierstra, arXiv:1509.02971 (DDPG paper): `https://arxiv.org/abs/1509.02971`
- *Proximal Policy Optimization Algorithms*, John Schulman, Filip Wolski, Prafulla Dhariwal, Alec Radford, Oleg Klimov, arXiv:1707.06347 (PPO paper): `https://arxiv.org/abs/1707.06347`
- TORCS: `http://torcs.sourceforge.net/`
- *Deep Reinforcement Learning Hands-On*, by *Maxim Lapan, Packt Publishing*: `https://www.packtpub.com/big-data-and-business-intelligence/deep-reinforcement-learning-hands`

Assessment

Chapter 1

1. A replay buffer is required for off-policy RL algorithms. We sample from the replay buffer a mini-batch of experiences and use it to train the $Q(s,a)$ state-value function in DQN and the actor's policy in a DDPG.

2. We discount rewards, as there is more uncertainty about the long-term performance of the agent. So, immediate rewards have a higher weight, a reward earned in the next time step has a relatively lower weight, a reward earned in the subsequent time step has an even lower weight, and so on.

3. The training of the agent will not be stable if $\gamma > 1$. The agent will fail to learn an optimal policy.

4. A model-based RL agent has the potential to perform well, but there is no guarantee that it will perform better than a model-free RL agent, as the model of the environment we are constructing need not always be a good one. It is also very hard to build an accurate enough model of the environment.

5. In deep RL, deep neural networks are used for the $Q(s,a)$ and the actor's policy (the latter is true in an Actor-Critic setting). In the traditional RL algorithms, a tabular $Q(s, a)$ is used but is not possible when the number of states is very large, as is usually the case in most problems.

Chapter 3

1. A replay buffer is used in DQN in order to store past experiences, sample a mini-batch of data from it, and use it to train the agent.
2. Target networks help in the stability of the training. This is achieved by keeping an additional neural network whose weights are updated using an exponential moving average of the weights of the main neural network. Alternatively, another approach that is also widely used is to copy the weights of the main neural network to the target network once every few thousand steps or so.
3. One frame as the state will not help in the Atari Breakout problem. This is because no temporal information is deducible from one frame only. For instance, in one frame alone, the direction of motion of the ball cannot be obtained. If, however, we stack up multiple frames, the velocity and acceleration of the ball can be ascertained.
4. L2 loss is known to overfit to outliers. Hence, the Huber loss is preferred, as it combines both L2 and L1 losses. See Wikipedia: `https://en.wikipedia.org/wiki/Huber_loss`.
5. RGB images can also be used. However, we will need extra weights for the first hidden layer of the neural network, as we now have three channels in each of the four frames in the state stack. This much finer detail for the state space is not required for Atari. However, RGB images can help in other applications, for example, in autonomous driving and/or robotics.

Chapter 4

1. DQN is known to overestimate the state-action value function, $Q(s,a)$. To overcome this, DDQN was introduced. DDQN has fewer problems than DQN regarding the overestimation of $Q(s,a)$.
2. Dueling network architecture has separate streams for the advantage function and the state-value function. These are then combined to obtain $Q(s,a)$. This branching out and then combining is observed to result in a more stable training of the RL agent.

3. **Prioritized Experience Replay** (PER) gives more importance to experience samples where the agent performs poorly, and so these samples are sampled more frequently than other samples where the agent performed well. By frequently using samples where the agent performed poorly, the agent is able to work on its weakness more often, and so PER speeds up the training.

4. In some computer games, such as Atari Breakout, the simulator has too many frames per second. If a separate action is sampled from the policy in each of these time steps, the state of the agent may not change enough in one time step, as it is too small. Hence, sticky actions are used where the same action is repeated over a finite but fixed number of time steps, say n, and the total reward accrued over these n time steps is used as the reward for the action performed. In these n time steps, the state of the agent has changed sufficiently enough to be able to evaluate the efficacy of the action taken. Too small a value for n can prevent the agent from learning a good policy; likewise, too large a value can also be a problem. You must choose the right number of time steps over which the same action is taken, and this depends on the simulator used.

Chapter 5

1. DDPG is an off-policy algorithm, as it uses a replay buffer.

2. In general, the same number of hidden layers and the number of neurons per hidden layer is used for the actor and the critic, but this is not required. Note that the output layer will be different for the actor and the critic, with the actor having the number of outputs equal to the number of actions; the critic will have only one output.

3. DDPG is used for continuous control, that is, when the actions are continuous and real-valued. Atari Breakout has discrete actions, and so DDPG is not suitable for Atari Breakout.

4. We use the `relu` activation function, and so the biases are initialized to small positive values so that they fire at the beginning of the training and allow gradients to back-propagate.

5. This is an exercise. See `https://gym.openai.com/envs/InvertedDoublePendulum-v2/`.

6. This is also an exercise. Notice what happens to the learning when the number of neurons is decreased in the first layer sequentially. In general, information bottlenecks are observed not only in an RL setting, but for any DL problem.

Chapter 6

1. **Asynchronous Advantage Actor-Critic Agents** (**A3C**) is an on-policy algorithm, as we do not use a replay buffer to sample data from. However, a temporary buffer is used to collect immediate samples, which are used to train once, after which the buffer is emptied.
2. The Shannon entropy term is used as a regularizer—the higher the entropy, the better the policy is.
3. When too many worker threads are used, the training can slow down and can crash, as memory is limited. If, however, you have access to a large cluster of processors, then using a large number of worker threads/processes helps.
4. Softmax is used in the policy network to obtain probabilities of different actions.
5. An advantage function is widely used, as it decreases the variance of the policy gradient. *Section 3* of the A3C paper (`https://arxiv.org/pdf/1602.01783.pdf`) has more regarding this.
6. This is an exercise.

Chapter 7

1. **Trust Region Policy Optimization** (**TRPO**) has an objective function and a constraint. It hence requires a second order optimization such as a conjugate gradient. SGD and Adam are not applicable in TRPO.
2. The entropy term helps in regularization. It allows the agent to explore more.
3. We clip the policy ratio to limit the amount by which one update step will change the policy. If this clipping parameter epsilon is large, the policy can change drastically in each update, which can result in a sub-optimal policy, as the agent's policy is noisier and has too many fluctuations.
4. The action is bounded between a negative and a positive value, and so the `tanh` activation function is used for `mu`. For sigma, the `softplus` is used as sigma and is always positive. The `tanh` function cannot be used for sigma, as `tanh` can result in negative values for sigma, which is meaningless!
5. Reward shaping generally helps with the training. But if it is done poorly, it will not help with the training. You must ensure that the reward shaping is done to keep the `reward` function dense as well as in appropriate ranges.
6. No, reward shaping is used only in the training.

Chapter 8

1. TORCS is a continuous control problem. DQN works only for discrete actions, and so it cannot be used in TORCS.

2. The initialization is another initialization strategy; you can also use a random uniform initialization with the `min` and `max` values of the range specified; another approach is to sample from a Gaussian with a zero mean and a specified sigma value. The interested reader must try these different initializers and compare the agent's performance.

3. The `abs()` function is used in the `reward` function, as we penalize lateral drift from the center equally on either side (left or right). The first term is the longitudinal speed, and so no `abs()` function is required.

4. The Gaussian noise added to the actions for exploration can be tapered down with episode count, and this can result in smoother driving. Surely, there are many other tricks you can do!

5. DDPG is off-policy, but **Proximal Policy Optimization** (**PPO**) is the on-policy RL algorithm. Hence, DDPG requires a replay buffer to store past-experience samples, but PPO does not require a reply buffer.

Other Books You May Enjoy

If you enjoyed this book, you may be interested in these other books by Packt:

Deep Reinforcement Learning Hands-On
Maxim Lapan

ISBN: 978-1-78883-424-7

- Understand the DL context of RL and implement complex DL models
- Learn the foundation of RL: Markov decision processes
- Evaluate RL methods including Cross-entropy, DQN, Actor-Critic, TRPO, PPO, DDPG, D4PG and others
- Discover how to deal with discrete and continuous action spaces in various environments
- Defeat Atari arcade games using the value iteration method

Python Reinforcement Learning Projects

Rajalingappaa Shanmugamani, Sean Saito, Et al

ISBN: 978-1-78899-161-2

- Train and evaluate neural networks built using TensorFlow for RL
- Use RL algorithms in Python and TensorFlow to solve CartPole balancing
- Create deep reinforcement learning algorithms to play Atari games
- Deploy RL algorithms using OpenAI Universe
- Develop an agent to chat with humans

Leave a review - let other readers know what you think

Please share your thoughts on this book with others by leaving a review on the site that you bought it from. If you purchased the book from Amazon, please leave us an honest review on this book's Amazon page. This is vital so that other potential readers can see and use your unbiased opinion to make purchasing decisions, we can understand what our customers think about our products, and our authors can see your feedback on the title that they have worked with Packt to create. It will only take a few minutes of your time, but is valuable to other potential customers, our authors, and Packt. Thank you!

Index

www.ingramcontent.com/pod-product-compliance
Lightning Source LLC
Chambersburg PA
CBHW080530060326
40690CB00022B/5081